LEAN
SAFETY

Transforming Your
Safety Culture with

ONE WEEK LOAN

LEAN SAFETY

Transforming Your Safety Culture with Lean Management

Robert B. Hafey

CRC Press
Taylor & Francis Group
Boca Raton   London   New York

CRC Press is an imprint of the
Taylor & Francis Group, an **informa** business

A PRODUCTIVITY PRESS BOOK

Productivity Press
Taylor & Francis Group
270 Madison Avenue
New York, NY 10016

Library of Congress Cataloging-in-Publication Data

Hafey, Robert B.
   Lean safety : transforming your safety program with lean management / Robert B. Hafey.
      p. cm.
   Includes bibliographical references and index.
   ISBN 978-1-4398-1642-4
   1. Manufacturing processes--Safety measures. 2. Industrial safety. 3. Industrial management. I. Title.

   TS183.H34 2010
   658.4'08--dc22   2009031079

Visit the Taylor & Francis Web site at
http://www.taylorandfrancis.com

and the Productivity Press Web site at
http://www.productivitypress.com

Dedication

To everyone who works with their hands and
attempts to stay out of harm's way. Stay safe.

Writing a book provided me the challenge of setting my own course—seeking personal growth rather than security. Without the unconditional love and support of my bride of many years, Sandy, this would have been an impossible task. Thank you for a lifetime of understanding.

Contents

Acknowledgments

Books, plant tours and the many contacts made while attending AME (Association for Manufacturing Excellence) events all contributed to building my lean knowledge base. AME is a volunteer organization that is truly unique. Only passionate volunteers can accomplish tasks on the scale of planning and executing a five-day lean conference with over 2,000 attendees. I am privileged to be an AME volunteer inasmuch as the rewards have exceeded my contributions. I have made good friends and learned a great deal about both lean and life from this group of selfless passionate volunteers. To each of you, I am grateful. The purity and longevity of this unique volunteer organization is dependent on a steady stream of practitioner volunteers. Don't pass up the opportunity of a lifetime to get involved.

During my 40+ years in manufacturing, I have gained, from many different people, a well-rounded understanding of safety in the workplace. I would like to recognize and thank all of them because, by sharing their manufacturing and safety knowledge with me, they have contributed to this book. When they read this book, they may recognize a phrase, sentence or concept because they may have taught that lesson to me.

Thank you to Mick Ramsey and Jerry Paulson, two unique and talented CEOs, who understand the only path to lasting continuous improvement is one built on trust and employee engagement. These lessons are a foundation piece of this book. Special thanks go to Bruce Paugh and Gary Mitas. Watching your passion and knowledge grow as we together built the "lean office" was a very rewarding experience. Keep the lean fires burning. To all the Gator Guards, past and present, for helping me better understand the many facets of workplace safety. Special thanks to Cheryl Jekiel and Michael Bremer, AME associates and friends, for sharing their lean knowledge and providing the spark to begin this project. Also, Michael Sinocchi, Lara Zoble, Jay Margolis, Maggie Mogck, and others at Taylor & Francis. Without their guidance and support, this would still be a stack of notes rather than a book. And, finally, a thank you to friends, family and co-workers whose interest in my writing was a source of continual encouragement. All of you, and many others I have not mentioned, are responsible for much of the content in this book.

Introduction

Writing a book was not on my life list until the fall of 2008. I was attending the 2008 AME (Association for Manufacturing Excellence) International Conference in Toronto. As a member of the conference planning team, I had, almost one year prior, proposed a new feature for the conference. It was the "Practitioner Idea Exchange Café." The concept came to me at the previous year's annual conference where I had volunteered to be a room host for a full-day workshop. My room host duties included making sure the presenter and the attendees were comfortable and had what they required. My reward for filling this role was a seat in the workshop. It turned out to be an extremely long day in which I watched almost everyone in the room suffering from a slow death by PowerPoint®. Just one presenter and seven hours of slides was putting the workshop attendees to sleep quicker than prescription sleep medication. There was no interactive component to the workshop until the seventh hour when the now thoroughly exhausted presenter sat down and said, "So, tell me about some of the continuous improvement activities in your plant." This one question was the equivalent of serving everyone a double shot of espresso. The room had come to life. The presenter with his one question had discovered a cure for death by PowerPoint. It was amazing how anxious the attendees were to tell their stories. There was only a half hour of scheduled time remaining, but they could have shared stories for hours. This was the moment I understood that all conference attendees at some level, not just the workshop consultants and the value stream practitioner presenters, had stories to tell, and the volunteer AME organization had to provide a forum for practitioner-to-practitioner sharing. Thus was born the "Practitioner Idea Exchange Café."

I had one year in which to develop my plans for the Café, so I would be ready to bring it to fruition at the 2008 AME Toronto conference. During that same year, I became involved in using kaizen blitz (multi-day, team-based rapid-improvement) events, which focused on safety improvement, in both our and other's manufacturing plants. It was a wildly successful, nonconventional use of a common lean tool. (Lean is a manufacturing philosophy that reduces the total cycle time between taking a customer order and the shipment by eliminating waste. Lean is based on the Toyota Production System (TPS). Since lean is a process review and improvement activity, it can be applied to all business processes, including safety, and all business types. Application of lean is taking place in medical, military, government, service, process, and manufacturing operations. If you are not familiar with Lean Management just think continual improvement while reading this book.) Safety and cycle time gains both resulted from the kaizen events in which no stopwatches or cycle time process step sheets were used. As a result of these successes, when I was planning the agenda for the seven facilitated sessions in the Café, I schedule an hour for "Using Lean Tools to Drive Safety Gains." Safety, to the best of my knowledge, had never been the sole topic of a session at an AME conference. Because I didn't know how it would be received, I asked one of my co-workers, who would be attending the conference, if he would partner with an ergonomist to do a 10-minute presentation on a safety kaizen event. The two of them had participated in a safety kaizen blitz just months before and I thought their story could be part of the scheduled Café safety session to ensure we had at least one topic to begin the discussion.

Exactly what I had witnessed at the end of the seven-hour workshop, one year before, happened in each of the seven sessions scheduled in the Café. I sat back, listened, and guided the conversations of 25 to 30 attendees in each of the seven sessions. Ideas and business cards were exchanged. The attendees brought and received value in each of the sessions with the safety session being no exception. As the safety kaizen blitz story was told, and then as other attendees shared their safety successes or posed questions asking for help, it was easy to recognize that safety was a value-added topic at a lean conference. Everyone in the room displayed a passion for and a desire to learn more about safety improvement. Lean and safety have always been interconnected. Any lean practitioner, who has invested time working with shop floor teams to reconfigure a work process, has witnessed the safety improvements that accompany the cycle time gains. Safety gains, however, are usually relegated to a second position behind the cycle time gains because the goal of lean is cycle time improvement. The positive results of the safety kaizen events I had facilitated earlier that year combined with my experience in the Café that day started me thinking about "lean safety," the safe path to lean, and a book on that topic. Instead of safety taking a back seat during lean events, why not bring it to the forefront? Here was a way to touch people's lives by improving safety in the workplace. Lean practitioners use lean tools to expose and then solve problems, so why not use lean tools and techniques to emphasize and improve safety?

Safety is a universal topic with a common language in every plant all over the world. A pinch point is a pinch point here and everywhere. Despite a common language and a general understanding of safety risks, safety programs vary greatly in scope and depth. The scope of a safety program is most often determined by the amount of safety legislation companies have to adhere to and the business leadership usually determines depth. Safety in this country is often touted as the "no. 1 priority," yet a great deal of the safety activity in companies is driven by compliance to legislation rather than proactive safety improvement programs. Therefore, not much depth is evident in many safety programs. Contrast safety management (a business requirement) with lean activities. Unless a lean edict has come down, the corporate pipeline lean is optional. No legislation requires lean activity. What usually drives lean in companies that have the option is a culture that needs repair. A serious lean effort will tear apart an old entitlement-riddled culture, with high costs and poor customer service, and build it into something new. Companies with ineffective safety programs that result in poor safety records also have a cultural problem. Present is a culture that does not value, or expect, a safe workplace and, therefore, people act accordingly. The common element is culture—it is the root cause of poor safety and ineffective businesses. It would then make sense that some common methods could be used to drive both the continual improvement of safety and general business processes. During the past 10 years, it would be a safe guess that at least 200 books with "lean," or some other continuous improvement-related title, have been published. Most everyone who works in operations could name at least a few of their favorite book titles. Yet, any one of these same operational managers, unless they have strong ties to the safety community, would be challenged to name one book title related to safety improvement. One of my hopes is that this book can unite the lean and safety communities by providing a convergence point.

Remember that buzzword from the mid-1990s—synergy? If a business was evaluating Enterprising Resource Planning (ERP) systems during that time, like our company was, they heard from each and every salesperson about the synergistic approach that was unique to only their software. Because I worked in operations, rather than software sales, to refresh my memory I looked up *synergy* on Wikipedia. *"Synergy (from the Greek synergos, συνεργός meaning working together) is the term used to describe a situation where different entities cooperate advantageously for a final outcome."* It has taken me over a decade to find a good practical example of something I could describe with the word synergy without feeling or sounding like a software salesperson. My example is lean and safety together, which is why, when trying to decide on a book proposal title, I chose "Lean Safety." It seemed appropriate because of the synergy I witnessed when using lean tools to positively impact safety.

Message for Leaders

Although this book touches on a mix of topics related to lean and safety, an underlying theme of employee engagement and ownership is core to both topics. People want to be treated with dignity and contribute to the success of the business in which they work. They want to feel as though they are in on things—involved and well informed about the business. When they are given that opportunity, they will provide discretionary effort that will yet again prove money is not the primary motivation for any human being. Only the leaders of a business can give their employees that opportunity.

When facilitating kaizen blitz events, I often use Post-it® notes to take quick-pulse surveys. At the end of a three-day safety focused kaizen, I asked everyone to write a note expressing how they felt about being given the "gift of time" to work on improving safety in their plant. One of the participants wrote, "Thank you for giving me the opportunity to learn." This message, though given to the facilitator, was intended for the leaders of the business. Leaders that must understand the requirement to guide, energize, and excite their employees to create the contagious commitment required for world-class safety or lean. Sustainability of world-class lean or safety is impossible without employee engagement and ownership. Only the management team can lead the development of a business culture that engages all employees in business improvement. Simply start by providing the "gift of time" for improvement activities and by demonstrating through your daily actions that safety is a priority. Leaders who follow this path can start creating their legacy today.

Message for Lean Thinkers

If you, like me, are infected with "lean fever," then you have a safety obligation. Your assignment, after reading this book, will be to use your facilitation skills and lean tools to PDCA (plan-do-check-act) your company's incident/accident investigation process. Engage and guide your company safety team using the A3 process to accomplish this task. Conclude with a future state map that includes leading all incident investigations yourself. Use your lean problem-solving skill set to get to root causes and corrective actions for every incident that occurs. Your lean knowledge and skills can be put to no better use than helping to ensure the safety of all who work in your plant. Once you experience the power of this combination (safety and lean), you will begin to understand how you can influence the acceptance of lean by following a safety path.

Message for Hourly Safety Team Members

You are the experts who often work in areas where the danger of injury is the greatest. Never take safety risks … never put productivity or customer needs ahead of your safety … stop and think safety before taking nonstandard actions … and guard against complacency when completing routine tasks. Become an uncompromising safety champion in your workplace. Help protect yourself and your co-workers by not settling for a safety program based solely on compliance. Compliance programs are important, but do not go far enough to prevent all injuries. Safety compliance is required work; safety improvement program activities are fun. Read this book and then hold both your management team and the internal lean staff accountable for helping you and your safety team develop a vision for a new safety program—a safety program based on the continuous improvement of safety. Then, have fun making a safety difference.

Brief Overview of Chapters

Chapter 1: Why Focus on Safety?

Chapter 1 challenges readers to put more emphasis on and accept individual responsibility for safety. A brief overview of the limits of a compliance-based,

OSHA (Occupational Safety and Health Administration)-mandated safety program is compared to a proactive safety program driven by lean thinking and tools. Learn about the value of a proactive program that engages employees in the continuous improvement of safety as seen from the eyes of a manager, a lean champion, and an hourly employee.

Chapter 2: Change the Culture

This chapter will reinforce the fact that lean is not a program, but rather a mindset or a way of thinking. Examples clarify why a successful lean implementation requires changing the culture of the business via employee engagement and involvement so that it is populated with lean thinkers that focus on customer and process cycle time reduction. This understanding combined with the adoption and use of lean tools can build a world-class safety culture. Another key point is that a lean leadership style and some common lean tools will allow you to improve both safety and lean at the same time.

Chapter 3: Leadership's Role

Leadership must be the first to change if the culture of a business is to move in a new direction. World-class lean or safety has to be built on a foundation of trust and the level of trust in any business is a direct reflection of leadership's behaviors and actions. Their responsibility is to focus everyone on business process improvement and to ensure that the support systems are in place, which will enable the changes to occur.

Chapter 4: Lean Tools for Safety

Engaging employees in continual improvement requires strong social skills, such as coaching, teambuilding, and facilitation. While using lean tools, which help develop their people skills, lean champions engage and grow employees because they recognize their success is dependent on these same people. The basic lean tools described are as applicable to safety as they are to lean and can be used to drive business improvement.

Chapter 5: Advanced Lean Tools for Safety

This chapter builds on some of the basic lean tools covered in Chapter 4 and the enhanced lean skill set developed while using them. It also covers the use of the A3 problem-solving process and a facilitated kaizen blitz—the most powerful people development and engagement tool in a lean thinker's toolbox. Both can be used to drive genuine, lasting safety improvement.

Chapter 6: Safety Program Leadership

World-class safety is dependent on engaging the workforce is proactive workplace safety improvement. No one person can guarantee a safe workplace. A distributed approach to safety leadership, using a formal, team-based structure, will be covered in this chapter. Developing a safety team of "passionate about safety" people provides both current safety leadership and a path for the development of future business leaders.

Chapter 7: Incident/Accident Investigation

Driving fear from the accident investigation process is a critical step in building trust. Lean leaders understand that every incident or accident is an opportunity to improve safety and build trust. They focus on the "what and why," and not the "who," while facilitating investigations. Their goal is to guide the investigation team to root causes and the identification and implementation of corrective actions that will prevent reoccurrences.

Chapter 8: Promoting Safety

Raising the level of safety awareness has always been the goal of progressive safety management philosophy. Rather than using safety banners and safety bingo to raise awareness of lean safety, the goal is to involve as many people as possible in the safety program. Proactive safety improvement activities that everyone can participate in are discussed in this chapter.

Chapter 9: Roadmap to World-Class Safety

The ability to impact people and, therefore, the cultural side of safety will determine the success of a safety program. In this chapter, the elements of an existing compliance-based safety program are contrasted with a pro-active, employee involvement-based, lean safety program.

Chapter 10: Safety Standard Work

Standards are the baseline from which all continuous improvement activities can be measured. By establishing standards for safety management and involvement, you will ensure that the audit step required to get to world-class safety is, indeed, in place. In this chapter, a list of possible standards for different positions are presented.

Chapter 11: Safety Metrics

What you measure makes a difference. Most safety programs measure only the negative metrics, such as OSHA recordable accidents, required for OSHA compliance. Lean safety programs measure the proactive, positive activity of employees engaged in safety improvement.

A few years ago a YouTube video clip, probably recorded with a cell phone, was being circulated in manufacturing circles. The plant where it had been filmed was in a country where safety regulations differ from what we would expect in the United States. Watching the video, one observed five individuals who, in between the strokes of a very large mechanical punch press, reached into the open die set simultaneously to transfer a part into or out of a die station. It was a human version of an automated mechanical transfer press. Many who viewed the video clip, and were familiar with OSHA's punch press safety regulations, were aghast at the inherent dangers of this action. Most people picked up on and made reference to the five individuals who were transferring the parts. I instead focused on the individual in the rear, to the right of the movie frame. Standing in a rather dark area, he was staring intently at the individuals feeding parts in and out of the die set. He was at the press controls and was responsible for initiating the stroke cycle once everyone had gotten their torsos and arms out of harm's way. At first I was sure he was not paid enough for the safety responsibility he had been given and accepted. Then, after watching this video clip a few more times, I realized this situation was not that unusual. Someone responsible for the safety of others is a common occurrence. I'm convinced that this is what individuals agree to when they accept a role in management. They may be a few steps removed from the press controls, but regardless, they are responsible for the safety of those who report to them directly or indirectly. Once acceptance of this safety responsibility is clear, the gravity of every unsafe situation or condition is highlighted. My hopes are that individuals use this book as a resource to marshal and combine the knowledge of their lean and safety staffs to reduce or eliminate unsafe situations and conditions. Then, after witnessing some early gains, they continue by engaging and involving every employee in building a world-class safety program based on the continual improvement of safety. If you follow this path, it will not lessen your safety responsibility, which never diminishes, but it will reduce the chance of serious injuries, which makes the "lean safety" trip worth taking.

A genuine "focus on safety" is about respect for people, while "safety first" is a saying printed on safety banners. Don't confuse the two.

Chapter 1

Why Focus on Safety?

Demonstrate That Safety Really Is First

I believe that safety really is first. Recently, I listened to a segment of an NPR (National Public Radio) program titled, *This I Believe*. I find this modern version of a 1950s radio program, originally created by Edward R. Murrow, very interesting because people seem to be talking from the heart. In creating *This I Believe*, Murrow said the program sought "to point to the common meeting grounds of beliefs, which is the essence of brotherhood and the floor of our civilization."[1] To recreate the 1950s program format, NPR invited anyone to submit a "this I believe" essay and then NPR selected poignant ones from the many entries. The selected essayists, some well known and some unknown, were recorded and asked to read their essays on air. The essay I listened to was written and read by the Olympian and heavyweight boxer Muhammad Ali and his wife Lonnie. It was a reflective look back on the values and beliefs that have guided him, and continue to guide him, along life's journey. While listening to that broadcast, the concept of sharing one's beliefs seemed like an appropriate way to begin a book about workplace safety. After all, with this book I hope to build a "common meeting ground of beliefs" regarding safety and the use of lean tools and management techniques.

This I believe: I believe safety is important. Each of us as human beings has a responsibility for our own safety and the safety of others. The first time we hold a newborn child we begin to truly understand safety responsibility. Once we strap a child into a car seat, we drive extra defensively to protect them from harm. It is easy to understand and accept our safety responsibility during adult and child interactions. An internal response triggers us to teach safety and protect those who seem vulnerable without our even thinking. I also believe that, if we say safety is the first priority in our workplaces, we should demonstrate that belief with action. Although some may believe adults "know better" and should be responsible for their own safety, I believe we have to teach and watch out for each other; visualize each other as children, so that we always instinctively think safety first. This I believe.

Workplace injuries hurt people and sometimes kill. In 2007, the U.S. Occupational Safety and Health Administration (OSHA) recorded 5,488 fatalities. After any serious injury occurs in a facility, the "safety first" banners are mentally unfurled and once again safety becomes "the priority" for everyone. Once again, because safety is just one of the many responsibilities of those who manage manufacturing plants, farms, construction companies, process plants, and other worksites where serious injuries can and do occur. Therefore, safety, identified as a top priority on most managers' long list of priorities, is often *not* given the focus it deserves. Then when adversity occurs, response is rapid, people pull together and, as a result, focused corrective actions happen. This is proved each and every time a workplace accident results in the loss of a life. Within the lean community, a comparison of this cavalier behavior is a similar lack of serious focus on lean implementations. Companies that take lean implementation seriously, from top to bottom, without the threat of imminent business failure, or some other very serious business hardship, are atypical. Many companies dabble in lean just as many companies give lip service to their safety programs. Lean and safety can become marketing programs in companies where leadership is more interested in advancing themselves than in helping others.

Lean is a current methodology business leaders can use to lead a revolution in their business. Businesses have been experiencing one revolution after another for the past 30 years. Examples include computers, integrated computer systems, MRP (material resource planning), automation, advanced machine tools, CAD (computer-aided design), robotics, lean and all of its earlier continuous improvement variations, and Six Sigma, just to name some. Most companies have taken advantage of many of the technology-driven opportunities to expand and grow their business. The business case, or ROI (return on investment) justification, for doing so, was a promise to increase or preserve profitability. Obviously the technology advances have provided a return. The often-publicized figures related to the reduction in the percentage of people employed in U.S. manufacturing are not just due to jobs being off-shored. Increased technology-driven productivity gains have played a major role. Why haven't continuous improvement programs, including the current favorite "lean," been as widely accepted and successfully implemented as these technology advances? Because lean is more than the continuous improvement of a business's process, it is equally about people and culture. Lean cannot be purchased like technology. Corporate leaders will quickly sign an ROI for a piece of proven technology that shows a favorable return. Little risk in this when compared to embarking on a culture-changing journey that requires many years of really hard work. Thus, for the implementation to be successful the business leaders have to be directly involved in leading the charge to change. Most opt to buy more technology.

In that same 30-year period, company safety programs have been relatively stagnant. The establishment of OSHA in 1970 was the last major force of safety program change.

According to Judson MacLaury (1984):

> The Occupational Safety and Health Act of 1970 heralded a new era in the history of public efforts to protect workers from harm on the job. This Act established for the first time a nationwide, federal program to protect almost the entire work force from job-related death, injury, and illness. Secretary of Labor James Hodgson, who had helped shape the law, termed it "the most significant legislative achievement" for workers in a decade. Hodgson's first step was to establish within the Labor Department, effective April 28, 1971, a special agency, the Occupational Safety and Health Administration (OSHA) to administer the Act. Building on the Bureau of Labor Standards as a nucleus, the new agency took on the difficult task of creating from scratch a program that would meet the legislative intent of the Act.[2]

Note that in the OSHA mission statement that follows they encourage continual process improvement in workplace safety.

> Employers are responsible for providing a safe and healthy workplace for their employees. OSHA's role is to assure the safety and health of America's workers by setting and enforcing standards; providing training, outreach, and education; establishing

partnerships; and encouraging continual process improvement in workplace safety and health.[3]

The opening sentence makes it very clear that employers, not OSHA, are responsible for safety. OSHA-mandated safety regulations both drove companies to comply and had a very positive impact on workplace safety. Results reported in 2008 show a positive trend.

> The Total Recordable and Days Away/Restricted case rates continued to decline, indicating that fewer American employees encountered safety or health hazards resulting in serious injuries or illnesses. The rates for calendar year 2007, reported on October 23, 2008, were lower than the previous year, and, thus, were the lowest rates that BLS (Bureau of Labor Statistics) has ever reported. Not only has the rate at which employees experienced a recordable injury decreased by 16.0% since calendar year 2003, but also the Days Away/Restricted case rate, the measure of cases in which employees were absent from work, restricted, or transferred as a result of a workplace injury or illness, has declined by 19.2% over the same period.[4]

The activities within many safety programs relate to or are guided by OSHA regulations. Or, to state it another way, businesses do what they are legislated to do. OSHA regulations have been good for everyone who is put in harm's way making the products and providing the services we require. This legislation has been responsible for the reduction of workplace injuries and deaths. The downside to this legislation-driven safety approach is that many company safety programs are reactive, with very little focus on safety improvement unless new regulations are mandated or a very serious injury occurs.

Management of safety within individual workplaces has for decades been a top-down directed activity. It mirrored the top-down management style in most businesses. Someone, or some functional group in every business with OSHA reporting requirements, has to maintain the required records and keep abreast of changes to the regulations. Safety, based on OSHA requirements, was pushed down the chain of command with the frontline supervisors as the last link in the safety management chain. Then, in the 1990s, the continuous improvement community began promoting teamwork and employee empowerment. A natural extension of those teambuilding efforts, in companies that engaged in this cultural changing activity, was safety teams. It was insightful and appropriate to engage the individuals most at risk to be injured to be part of safety management. The low number of companies that successfully built a teamwork culture during the 1990s mirrors the implementation of lean results today, and for the same reason. Teams were tools for changing the culture and most leaders did not have the stamina for the lengthy journey required to change their business cultures. Most companies, as a result, maintained their top-down management approach to safety management. Missing in most all safety programs, team-based or management-directed, is a methodology to proactively improve workplace safety. Compliance to regulations and safety audits, along with the corrective actions resulting from injuries, drive most safety program activity today. Despite what many may think, hanging safety banners and playing safety bingo are not proactive safety activities. They are questionable safety awareness activities at best. To realize behavioral and workplace physical safety changes, in addition to heightened safety awareness, you must overhaul your current, outdated safety program.

This new safety program renovation methodology, "lean safety," is a model based on employee engagement and business improvement using the tools in a lean thinker's toolbox. Lean safety is a self-serving activity. Those engaged in proactive safety improvement will be reducing injury risks related to the very work they perform. Hourly wage earners, as well as management, need to know what's in it for them if they are to become truly engaged in this business-changing safety effort.

Why, if management has struggled to implement lean, would they want to embark on an employee engagement effort intended to revamp the company's safety program? For multiple reasons, but the first and most important is because preventing injuries is the right thing to do. Anyone familiar with lean will be familiar with the Toyota Production System (TPS). Lean is the TPS. All of the lean tools used to drive continual improvement at Toyota now fill a lean thinker's toolbox. But the tools are only half of the lean story at Toyota. Equal to ongoing continual improvement is the other philosophical component of TPS—respect for people. How better to show respect for people than engaging them in making their workday safer. That is a genuine activity, and it is the right thing to do. We can be driven to action by a serious injury or we can do the right thing. I once worked at a facility where a fatality occurred. It is something I will never forget and the impact on a co-worker was even greater. He supervised the deceased and discovered the fatality when he was making his rounds. That situation would cause anyone to spend the balance of his lifetime asking himself a very difficult question: Could I have prevented the accident had I done more to ensure safety was first? Focusing on safety is the right thing to do because it will help to prevent workplace injuries—minor and severe. Management needs to demonstrate that safety is first before they are driven to action by reacting to serious injuries. Clearly, in a world-class safety facility, safety would be more than a priority; it would be one of the corporate values.

The second reason management should consider a rework of their safety program is for financial reasons. A lack of safety and the resulting injuries are costly. To protect businesses against costly litigation, when injuries occur, they carry state-mandated worker's compensation insurance. Worker's compensation insurance costs may be equal to 2.5 percent of the average employee's wages.[5]

A company's injury history, both the frequency and severity of injuries, is one of the factors determining their workers compensation insurance premiums. Having a revamped safety program will drive down injury rates and, in turn, insurance premiums. But, that is only one of the costs associated with workplace injuries. Others include medical expenses for treatment of the injured and lost productivity. Below are some rather startling findings from a PBS *Frontline* program based on a University of Michigan Press publication, which examined workplace injury statistics for 1992.

- Roughly 6,371 job-related injury deaths, 13.3 million nonfatal injuries, 60,300 disease deaths, and 1,184,000 illnesses occurred in the U.S. workplace in 1992.
- The total direct and indirect costs associated with these injuries and illnesses were estimated to be $155.5 billion, or nearly 3 percent of gross domestic product (GDP).
- Direct costs included medical expenses for hospitals, physicians, and drugs, as well as health insurance administration costs, and were estimated to be $51.8 billion.
- The indirect costs included loss of wages, costs of fringe benefits, and loss of home production (e.g., child care provided by parent and home repairs), as well as employer retraining and workplace disruption costs, and were estimated to be $103.7 billion.[6]

Despite the fact that this data is from 1992, and we all know medical costs have increased substantially, the facts noted above are staggering. Spending a company's resources to improve safety by switching a compliance-based safety program to one that proactively improves safety can help to reduce all of these direct and related indirect costs.

The third reason why management should jump on the lean safety band-wagon is because focusing on safety, using lean tools, will jump-start stalled lean efforts. If a trust gap is preventing lean from gaining momentum, build a safety bridge to span the void. Start building trust through the genuine efforts of safety improvement. In future chapters, I will review safety focused continual improvement activities that resulted in the compound benefit of both cycle time and safety gains. You can have both by focusing on one or the other. Lean and safety are interlocked. Safety is the easier route to success.

Why, should an hourly worker support, and get involved in, the rebuilding of their company's safety program? For the same reason—it is the right thing to do. When management offers the "gift of time" to improve workplace safety, people should jump at the opportunity. It is a chance to expand their knowledge base and become a more valuable employee in this age of life-long learning. Learning can be energizing and rewarding if individuals allow themselves to be pushed beyond their self-imposed comfort zone. Hourly workers are the experts in the areas where the greatest safety hazards exist. They can step forward and make a difference by contributing their expertise to improving workplace safety. When they do, any uncertainty about the value of lean will dissipate as they participate in activities using lean tools to drive safety gains. They will begin to understand the value of lean tools because a safer work environment will result from their personal efforts. They and their co-workers will not only be safer, their work will be easier. Easier because the same lean tools used to advance safety will have reduced the cycle times of the work they perform. This lean safety activity, which engages both salaried and hourly staff, will improve safety and cycle times and help to establish a continuous improvement culture. The continuous improvement of work processes and safety can be the same activity.

Quick Guide: Lean Focused Approach to Safety Management

- "Lean safety" is a change management tool that requires direct management involvement.
- OSHA compliance-based safety programs are lacking a continuous improvement component.

- OSHA's mission statement clearly states that employers are responsible for providing a safe and healthy workplace.
- OSHA encourages continual improvement in workplace safety.
- How better to show respect for people than engaging them in making their workday safer.
- Worker's compensation insurance costs may be equal to 2.5 percent of the average employee's wages.
- Spending a company's resources to improve safety by switching a compliance-based safety program to one that proactively improves safety can reduce all of the direct and related indirect safety costs.
- If a trust gap is preventing lean from gaining momentum, build a safety bridge to span the void using safety as the path to lean.
- Hourly employees are provided a chance to expand their knowledge base and become a more valuable employee in this age of lifelong learning.
- "Lean safety" activity, that which engages both salaried and hourly staff, will improve safety and cycle times and help to establish a continuous improvement culture.

Endnotes

1. National Public Radio (2005) *The history of 'This I Believe.'* http://www.npr.org/thisibelieve/about.html
2. MacLaury, J. (1984), *A history of its first thirteen years, 1971–1984*. OSHA, United States Department of Labor, Washington, D.C. http://www.dol.gov/oasam/programs/history/mono-osha13introtoc.htm
3. United States Department of Labor (2008) OSHA's role. Washington, D.C. http://www.osha.gov/oshinfo/mission.html
4. United States Department of Labor (2008) *Injury and illness rates: Record lows in FY2007*. Washington, D.C. http://www.osha.gov/as/opa/2008EnforcememtData120808.html
5. Welch, E.M. (2008) Workers' Compensation Cost Data, Michigan State University, East Lansing. http://www.lir.msu.edu/wcc/documents/WCData.pdf
6. Leigh, J.P., Markowitz, S., Fahs, M., and Landrigan, P. (2000) *Costs of occupational injuries and illnesses*. Ann Arbor, MI: The University of Michigan Press.

Chapter 2

Change the Culture

A Common Goal of Both Lean Philosophy and Safety Programs

Much has been written about manufacturing companies that have failed at lean implementations because they have only applied some of the lean tools and neglected to address the people or cultural side of their lean implementation. A natural starting point for businesses, when beginning a lean journey, is the application of some foundation-type lean tool, such as 5S (sort, set in order, shine, standardize, sustain), which is a simple workplace organization concept that everyone can understand and apply in their workspace. However, applying 5S and other lean tools does not make a company lean—far from it. The real goal of lean is not just the implementation of lean tools because that can be, and often is, perceived by employees as some sort of program. Lean is not a program, it is a way of thinking and seeing the world. It is a mindset. Individuals who are lean thinkers see businesses and the world as a collection of processes that can be taken apart and continually improved. Once you become a lean thinker, it is almost impossible to turn it off. No matter where you are or what you are doing, you are observant of opportunities to improve processes.

An example to help one understand this inability to turn off lean thinking occurred a few years ago while I was on a domestic flight. It had to be a few years ago because this incident is about food service. In this case, the flight attendant was working her way down the aisle distributing box lunches.

The problem I observed was that the box lunches were tightly packed four to an overwrap carton. Each time the attendant opened the lid on a new overwrap carton she struggled to get her fingers around one of the box lunches to extract it, using an extended pinch grip, from the carton. Once the first one was extracted with much physical effort, she could then easily remove the other three. When she approached me, I commented on her difficulty and offered a process change that would eliminate the ergonomic-related injury risk associated with the work process. I suggested that after she opened the overwrap cartons she should simply turn them upside down to slide the four box lunches onto her cart top. Later when she was collecting the empty lunch boxes from passengers near me, she stopped by to tell me what a great idea I had given her. She noted that she used the new technique for serving the balance of the lunches and that it was so much easier. The suggestion obviously made her job safer, but did it lead to real change? Did her airline have a culture that both supported her improvement ideas and then her involvement in implementing the work process change? Did all flight attendants then adopt this new method that would make their jobs safer and faster? Obviously I will never know. But, I do know that this is the real goal of lean. True lean is based on changing the culture of the business via employee engagement and involvement, so that it is populated with lean thinkers that focus on customer and process cycle time reduction. If everyone on that airplane had been observing the attendant's difficulty and offered suggestions to improve the lunch box extraction process, just as I had done, a small lean culture would have been present on that plane. "Not likely to happen," some might be thinking and they are right. Most passengers were more likely thinking that it was taking a long time for food service. I think "lean" and am passionate about it. I cannot switch it off. The others on the plane, just like many in your plants and offices, are disengaged and waiting for the plane ride or their shift to end. Therefore, the challenge confronting leaders is how to engage everyone in lean or safety improvement efforts. One of the main points of this book is to make it crystal clear that the path to either lean or safety improvement is the same. A common style of leadership and some common tools will allow you to improve safety and lean at the same time. Lean and safety are inextricably linked like pasta is to the Italian culture. Or, like the title of this book: lean safety.

Companies attempting to build world-class safety programs have long recognized the need to build a safety culture. I chose the word *build*. You can substitute *grow*, *develop*, *establish*, *create*, or another word of your choice to put in front of culture. The word that is really important is *culture*. It is not the elements of your safety program alone that will get you to world-class safety, it is when you impact the culture so that all individuals, without exception, make safety their daily priority. That can only occur in a culture where "safety first" is not just a slogan on a banner, but where it is a deep-seated reaction, just like looking both ways before crossing a street. Thus, as you can see, a common requirement to build, grow, or create world-class lean or world-class safety is to impact the culture within your organization.

How do you describe the culture of a business? Somewhere I heard the following description, which is the one to which I most often refer. It is how the individuals who work in the business think, act, and interact. You can walk into a business and very quickly know if safety or lean is cultural because it can be observed in the actions, interactions, and communications that take place. That personality of the place, good or bad, is evident to all who visit and observe. I have walked through many different manufacturing plants in my career and have observed cultures in which the employees feel valued and engaged in the work they perform and through others where people obviously cannot wait to leave at the end of their shift. Why is it that in some plants the employees do not even make eye contact as you pass and in others they smile, greet you, and seem genuinely happy you are there? Why the difference? Both are adult workplaces, but, with certainty, the level of trust in the two facilities will be quite different. Moving a culture toward one where employees feel engaged requires a growth in the level of trust. I intentionally used the term growth, for the trust level in any business will wax and wane based on many factors. However, if everyone agrees that businesses take on the personality of their leaders, then leadership must own the responsibility for the culture in their business and, in turn, the level of trust present or absent.

Therefore, the "right" culture for moving toward and attaining world-class lean or world-class safety is one in which there is a high level of trust. Sounds simple, but how is trust established? I believe it is with one person at a time. As trust is built with individuals, you can watch their motivation increase. It takes motivated people to get to world-class anything and changes in employee motivation are gradual or almost organic. There is no switch you can throw that will move them from formerly demotivated to now highly motivated. People respond to the extension of trust just as plants respond to sunlight and fertilizer. Individuals who have worked hard to develop high levels of trust in their organizations understand the importance of trust. These effective leaders understand that you cannot manage change—you must lead it. I clearly remember when a member of our executive team returned from a workshop held at a local plant that had embarked on building a continuous improvement team-based culture. Upon his return, he stopped by my office to provide an overview and noted that he had had an "ah-ha" moment while there. "You earn trust by giving it," he said. He stated it with the understanding that it was a very simple statement, yet one of deep insight. It impacted him and influenced his actions for years to come. He toiled tirelessly to build trust in the organization he led by extending trust to everyone. You earn trust by giving it. That is something that not only has to be understood, but personally practiced if the goal is a world-class safety or lean culture.

When people are engaged, the opportunity to build trust exists. Much of this book will be devoted to the use of lean tools to engage people in safety improvement. What exactly does engaging someone mean? I could give you the definition from a dictionary or Wikipedia, but what helped me understand it best was a discussion with one of my staff. This individual worked in one of the production support departments in a technical role, but had been given the opportunity to participate on a lean team that was assessing the work practices in one of our cells with the objective of reducing ergonomic injury risks. As is often the case with people in a production leadership position, I was also the champion of continuous improvement and, therefore, was leading the lean effort in which he was involved. When that project ended, I went to his supervisor and reported on what a great job he had done. He had made a real difference, so I asked if he could be freed up from his regular duties so that he could devote half of his week to continuous improvement. An agreement was reached and then within a year he was assigned to me full-time to do nothing but support our continuous improvement efforts. His new role lacked a formal structure and I gave him little supervision. Instead, I extended trust and allowed him to support people in their process improvement efforts. He proved that given a chance in this new role he could add real value to our organization. When the time came for his annual review, we sat down to talk about this new role and his performance. When I asked how he felt about being in this role compared to his past role in the business, he stated he really was enjoying the change. When I asked why he said, "Because I'm engaged." I then asked what he meant by that and he said, "You know, I am making a difference— I'm involved. I'm having fun helping people and improving the business." Just imagine a business culture where the majority, or even many, of the employees felt that way about their jobs. Everyone's best day at work is one in which they were involved and made a difference. If anyone reflects back and thinks about his/her best days at work, I think he/she will agree. How individuals can provide the opportunity for others to make a difference, be engaged in, and positively impact safety, is what I want to address in this book. But, before we get into the chapters on engaging people in safety process evaluation and improvement, let's explore leadership's role in culture building. More specifically, I will focus on a lean leadership style.

Quick Guide: Change the Culture

- Lean is not a program, but rather a mindset or a way of thinking.
- Successful lean implementation requires changing the culture of the business via employee engagement and involvement so that it is populated with lean thinkers who focus on customer and process cycle time reduction.
- A lean leadership style and some common lean tools will allow one to improve safety and lean at the same time.
- It is not the elements of one's safety program that will get one to world-class safety, it is when you impact the culture so that every individual, without exception, makes safety their daily priority.
- The culture of a business is how the individuals who work in the business think, act, and interact.
- Moving a culture toward one where employees feel engaged requires a growth in the level of trust.
- Management owns the responsibility for setting the culture in their business and, in turn, the level of trust present or absent.
- You earn trust by giving it.

Chapter 3

Leadership's Role

Focus Everyone on the Process

This is *not* another book on leadership. There are many qualified writers who have explored that philosophical subject. I recently read a blog posting in which the blogger had some fun referring to the stacks of books written on the topic of leadership. He noted that one could explore the leadership styles of numerous corporate and military leaders, Jesus, and hundreds of other lesser-known examples. Certainly, if someone read them all, or even some of them, he or she could distill leadership down into some core principles that, if followed, would make him or her a better leader. I am sure that that book has also been written—and more than once. Almost all leadership books have a person of power or fame positioned as the leader whose leadership style is explored in the book. That is not what I want to explore. More relevant to this book, and the topic of lean and safety leadership, are examples of leadership situations rather than the principles or philosophy of any particular leader. Personal change may be the first responsibility of leaders who begin the lean journey. It is a necessity for them to change from focusing on people as a cost to focusing on engaging people, as an asset, in process change. Lean leaders understand this empowering leadership style, not because they have read about it, but because they gained the understanding through their hands-on experiences.

When I talk about leadership I am not only talking about the senior leaders of an organization. Leadership can be and often is provided by people at all levels of an organization. A common trait of all leaders is the ability to accomplish goals through their people. Your reports will make you look great if you understand, and then demonstrate through your actions, that you are dependent upon them just as much as they are dependent upon you. Leadership never feels better than when you witness the honest work, results, and growth of those you lead.

A few years ago, I had the opportunity to facilitate a kaizen blitz (rapid continuous improvement) event as an AME (Association for Manufacturing Excellence) volunteer. Team-based kaizen events are the perfect training method to help leaders understand lean leadership. This event was not at the manufacturing plant in which I worked, but in one that had volunteered to host this three-day kaizen event. It was one of my first experiences at facilitation of a kaizen blitz and having that responsibility in someone else's plant really accelerated my learning curve.

Managers are paid to make decisions, so when approached with a problem, they usually respond with a solution. Stepping into this unfamiliar plant prevented me from providing ready solutions. I was forced to fill the role of a facilitator. Facilitators have to have the right questions, not the right answers. They are only to guide the team toward a successful outcome; they do not deliver the results. Being in an unfamiliar plant with people I had just met put me into the perfect learning position of having to rely on others for my success. I had just three days to provide lean leadership and earn the team's trust by respectfully engaging them in process analysis and improvement.

The team was a cross-functional group in that it was composed of both managers and hourly employees from this site and elsewhere. Because this was a publicly advertised AME event, there were external participants from different industries and professions. The agreed upon team charter spelled out the team's main objective, which was to reduce the changeover time on a three-ring binder machine. This equipment, when loaded with the correct components, assembled them into a standard three-ring binder. When selecting people to staff a kaizen team, the process expert is always included as a team member. That is the individual who performs the work process that the team will observe and improve. In this case, it was the three-ring binder machine operator. He was a Hispanic gentleman whose English, though better than my Spanish, was limited.

Day 1 of the event began with some overview training for the team. PowerPoint® slides were used to spell out the team guidelines, their objective, and the agenda for the next three days. Buzzwords and phrases were plentiful: "creativity before capital," "there are no bad ideas," "the first idea is never the best idea," "respectful employee engagement," and "have fun" were just a few used to prep the team. As the morning of training proceeded, I observed that the machine operator seemed to feel a little out of place. I am sure he was not accustomed to sitting in a meeting room listening to training material with a lot of unfamiliar people. His area of expertise and his comfort zone were both on the shop floor, which is where the team headed when the training had concluded. They were going to observe the "current state" machine changeover process they were charged with improving.

When the operator was told to begin, a stopwatch was started so the team could establish their baseline changeover time for the "current state." Now very comfortable, the operator moved efficiently from step to step of the changeover. As he did, all team members were challenged to identify any possible improvements. Because this machine could produce many different sizes of binders, the changeover consisted of many adjustments to guides, rails, and feeding systems that brought in the components to the assembly stations. The team quickly observed that the operator had to physically struggle to move some of the rotating and sliding mechanisms in order to make his adjustments. The team honed in on these steps of the changeover and, as they did, I observed the operator's interest and involvement grow. Those difficult elements of the changeover were noted on the team's opportunity log and over the next two days the team, and maintenance personnel from the host site, worked hard to improve these changeover steps that were both physically taxing and frustrating for the operator.

On the afternoon of the third day of the event, the team again observed the same changeover: "the future state." All the improvements had been completed, the operator had been briefed on the new "standard work" method for the changeover and he performed his work tasks smoothly and with much less effort. The team attained their goal of a 50 percent improvement in changeover cycle time. High fives and group hugs followed as all felt good about their combined effort and goal attainment. Because the team felt a sense of accomplishment, so did I. All of us external participants, soon after the event wrap-up meeting and celebration, said goodbye to our hosts and went back to our regular jobs without knowing the long-term impact of the changes that were made. Kaizen events are often criticized because the changes made do not stick. People sometimes go back to their old way of doing things. Did that happen in this case? It was well over a year later when I received an answer to that question.

I was back in the same plant for a half-day lean workshop presented by their senior lean manager. As part of the event, a plant tour was conducted.

As we were passing the binder assembly machines, the operator who had been on our kaizen blitz team recognized me and literally ran out to the aisle to meet and greet me. He said the machine was running great and he thanked me again and again for making his job so much easier. I was humbled because it was the team who really deserved his feedback. However, what he did for me was to reinforce an important lean leadership lesson— the process was the problem, not the operator. Machine operators do not come to work to do a bad job (or to get injured), but they often have to work in bad processes, which yield poor results. Those results are often the basis of the feedback they receive on their job performance. By focusing the kaizen team's efforts on the machine operator's process problems, they made his work both safer and easier to perform and, as a result, helped him become a more productive employee. It was at this time that I started to understand the strong connection between these lean events yielding both productivity and safety gains. This operator no longer had to physically struggle to make his machine changeover adjustments. The kaizen team had reduced both injury risks and cycle times.

The concept of focusing on the process, like most lean philosophy, is very simple to understand. Lean tools and techniques all seem to be based on a commonsense approach. So, what prevents the wide acceptance, or more importantly, the success of lean in more businesses? There is no one answer, but here are a few I would ask you to think about.

- Management that is ever changing
- Managers that are grounded in traditional thinking
- Managers that see their employees as only a cost and not as an asset

A change or turnover in the senior leadership of a business makes it very difficult to establish a lean culture because there is no ongoing constancy of purpose for the business culture. Some large businesses make it a developmental practice to continually and frequently change leaders. That practice may benefit the individuals personally and provide for leadership succession, but it is a detriment to lean-centered improvement because the level of passion for lean varies based on the leader's understanding. Yet, every leader understands his or her responsibility when it comes to safety. By using lean tools to drive safety improvement, as will be presented in the body of this book, one can demonstrate to leadership the power of lean.

Figure 3.1 is a comparison of someone with a traditional focus with that of a lean-thinking, process-focused individual. To help you interpret

Traditional Focus	Process Focus
Employee is the problem	Process is the problem
Doing my job	Helping to get things done
Understanding my job	How my job fits into the total process
Measure individuals	Measure processes
Change the person	Change the process
Can find a better employee	Can always improve the process
Motivate people	Remove barriers
Control employees	Develop people
Distrust	We are in it together
Who made the error	What allowed the error to occur
Correct errors	Reduce variation
Bottom-line driven	Customer driven

Figure 3.1   Traditional versus process-focused leadership.

the chart, read the following comparison based on the two columns on the chart. Rather than seeing employees as the problem and asking them to only do and understand their job, you help them understand that the process is the problem. You ask them to help get things done in order for them to start to see how their job fits into the total business process. Instead of measuring individuals and trying to change them by threatening to replace them, change and improve the business process in which they are involved. You quickly understand that rather than trying to motivate and control people, which leads to distrust, you remove barriers, develop people, and start building a sense of "we are in this together." When errors occur instead of spending time trying to find out who made the error, and then correcting the error, you pull your people together to talk about what allowed the error to occur and seek solutions to reduce variation. And finally focus everyone on the customer rather than the bottom line.

Key to understanding this lean leadership style is the acceptance of the fact that the process is the problem, not the person. Managers who are process-focused leaders and mirror the actions of the lean thinker described in the paragraph above see their staff as an asset and not just a cost. Once this leadership style is accepted, the blame game ends and the building of trust can begin. Company cultures, lean or safety, are built on a foundation of trust.

What role does process-focused leadership and trust building play in developing a world-class safety culture? I have worked for both a very large corporation that had a full-time safety director and a small- to medium-sized firm in which many people shared in the responsibility of safety leadership. Any manager with safety leadership responsibility, no matter what the size of the company, should understand that people do not care how much you know about safety until they know how much you care about their safety. If workers trust that management is looking out for their safety, it is an outcome of management's actions regarding safety.

Both what leaders say and how they say it is very important. You can disengage your employees simply by making statements like, "Let's get out there and work harder." When a statement like that is made, it implies that people are there to use only their hands. Contrast that with a lean leader who might encourage everyone to find ways to improve their work processes in order to make the work easier, safer, and, therefore, more productive. His request is asking them to use their brains as well as their hands. People listen very attentively when their leaders talk. I learned that lesson some years ago. At that time our business had just reached a safety milestone. We had attained one year with no lost time accidents. An all-company meeting was called as a way of celebrating that accomplishment. Our COO (chief operating officer) took the podium to address everyone and began by asking a simple question: "If you believe this is a safe place to work, raise your hand." I believe most everyone, including myself, raised our hands. After all, we were flush with the good feeling of making the one-year goal and it must be a safe place if we had attained that goal. He quickly set us straight. He spoke with passion and the utmost seriousness when he said very emphatically, "No, this is not a safe place to work; as a matter of fact, it is a very dangerous place to work." He went on to point out that our business operations had many risks and hazards that could seriously injury someone. He continued by emphasizing that it is only constant vigilance and high safety awareness that kept us all safe. He couldn't have been more serious. He used what was to be a celebration meeting to send a brilliant leadership message. Everyone listened and I will never forget the lesson. He understood and was sending the message that no one person can be responsible for safety within any plant. It was everyone's responsibility. Like most of you, he understood that increased safety awareness was critical to an effective safety program. Individuals must take ownership of their own safety. Companies can provide safety equipment, training, etc., but in the end it is the individuals who "thinks safety" before they act that create a good or bad safety record in a company. Our COO understood that only the combined safety efforts of everyone had gotten us to one year without a lost time accident and he wanted to use this opportunity to reinforce that message.

So when large corporations give someone the title of Safety Director, it is almost as if they are setting them up for eventual failure. It appears to me that leadership is too busy to be directly involved with safety, so they create a position to be responsible for it. Not a job I would want. Safety directors must spend countless hours worrying about the negative metrics, such as lost work-days, used to judge their performance. If they take their safety role seriously, they must suffer sleepless nights. If they aren't serious about the position, they probably rely on the blame game when injuries occur. It must be the fault of the individual who was hurt because he/she was responsible for his/her own safety, right? You can bet the words "unsafe act" probably appears in most of the safety directors' accident investigation documents. Now, contrast that with a lean leadership style that puts focus on the process, versus the individual, when problem solving. A lean leader focuses on the "what and why," not the "who." It is not a blame game, but a process review and improvement effort. Since a business is nothing but processes, a company's current safety program can be broken down into safety-related processes. Lean leaders using lean tools can engage employees in evaluating and improving each of these safety processes, which together constitute the safety program. My recommendation, if you are going to give one individual the responsibility for safety, is to give them the title of Director or Manager of Safety Improvement, then expect just that—a focus on improvement along with compliance.

The whole idea of a business cultural change has a strong connection to process evaluation and change. The employee engagement tool used in the company where I worked was teamwork. Everyone was on a team and was engaged in team-building activities. Teams worked in customer-focused cells, had weekly meetings, and maintained their own performance metrics. This team building effort changed our company culture. However, to accomplish this, the processes that were in place, which helped to define our "current state" culture, had to be changed to establish the "future state" culture. The reason they have to change is to drive the expected behaviors of the new culture. To initiate the cultural shift from a top-down management structure to a team-based structure, a "vision for the culture" document was created. It spelled out for everyone the expected behaviors in the new culture. It contained words, such as *learning organization, empowered employees,*

Areas of Change	Methods Employed
Leadership	Supervisors became coaches
Empowerment	Customer requirement information provided to the work teams for decision making
Education	Provided during work hours. Ranged from teambuilding and communications to computer skills
Teamwork	Teams were structured and managed with a four phase development process
Continual improvement	All employees required to submit process improvements
Annual review process	Program documents modified to include new expected behaviors
Metrics	Teams monitored and reported on their performance in six categories at quarterly meetings with management
Communications	Business financial information shared with everyone

Figure 3.2   Processes employed to change a company culture.

customer focused, motivated, adaptable, highly involved, individually accountable, and *working cooperatively.* Years of effort by many dedicated people slowly moved the culture. Figure 3.2 illustrates some of the areas of desired change and the methods deployed.

The take away from this example of a company using teamwork to engage all employees is that a work culture will only change if you modify some of the current processes that drive your existing culture.

A lean leader also understands that processes must have owners, and support systems have to be in place to help ensure process stability and improvement. For instance, in the example from Figure 3.2, someone had to take ownership of the annual review process and documents. To understand the steps or evolution of a business process, examine Figure 3.3.

The all-important Level 1 in Figure 3.3 spells out the requirement to identify the process owner. If this first step has not been completed, the process will often be out of control. An example that I have witnessed concerns forklifts. Many departments have forklifts and forklift operators. Someone is responsible for calling service when a forklift breaks down. The training manager had taken partial ownership of selecting a forklift operator training program. One of the existing forklift operators trained new forklift drivers. Daily forklift inspection documents were completed and filed away by a

Level 1	Process identified and process owner selected
Level 2	Process has been documented
Level 3	Process standardized and in use
Level 4	Process measured, evaluated and improved
Level 5	Process is showing positive trends

Figure 3.3   Illustration of business process growth progression.

clerk. The safety committee investigated the forklift incidents and accidents. This distributed responsibility method of managing a forklift program is missing one important ingredient: a process owner. Therefore, the forklift process will not move to Level 2 on the process evolution step chart and will remain a process out of control. If someone accepted ownership of the aforementioned forklift processes, they would ensure that the training process was evaluated, forklift repairs were tracked to justify purchasing replacements, different maintenance service providers were evaluated, all forklift accident and incident investigations were completed, forklift inspection forms turned in daily, and new forklift technology was understood. When no owner is defined, everyone involved does their best to keep the wheels from falling off. But, it is an uncoordinated process wreck. Why does this happen? Because most businesses are functionally organized, they have individually defined departments, such as engineering, purchasing, planning, accounting, and many others. Notice there is no forklift department, thus most department heads do not even see forklifts as a process that should have an owner, and even if they do, they do not want to add it to their list of responsibilities. It doesn't fit neatly into their or the company's departmental structure. So, it should be obvious that someone already in a leadership position has to add forklifts to their list of responsibilities. It may be the maintenance manager, the operations manager, or the plant manager who is responsible for the forklift process. Whoever it is must provide the individual leadership and a clear understanding of the requirement to take the forklift process through the five levels shown in Figure 3.3.

Safety is a process that intersects almost every departmental boundary. It may have an assigned process owner, like the safety director I mentioned above, or it may be and often is a distributed responsibility shared by many. However, final responsibility for safety goes way to the top of the organizational chart. Ask those business leaders responsible for leading people and they will tell you their no. 1 responsibility is the safety of their people. It is what they have to say. And just because they said it doesn't mean they are, or should be, directly involved in the safety processes. That often depends on the size of the company. A common responsibility, no matter what the size of the company, is to ensure the resources are available so that the safety processes are in control. This book is not for those leaders who say "safety first" because it is expected. It is for the leaders who want to be involved in safety because they want to impact safety and provide a safe workplace. These leaders, from plant managers to frontline supervisors or safety committee members, can make a real difference. By using a lean, process-focused leadership style along with some simple to understand lean tools, they can impact the safety culture in their businesses by engaging others in safety process definition and improvement. As they do this, they will positively impact the safety culture one person at a time.

Quick Guide: Leadership's Role

- One of the first responsibilities of leaders who begin the lean journey is to change, change from focusing on people as a cost to focusing on engaging people, as an asset, in process change.
- Leadership never feels better than when you witness the honest work and results of those you lead.
- Facilitators have to have the right questions, not the right answers.
- The process is usually the problem, not the person.
- Machine operators do not come to work to do a bad job, but they often have to work in bad processes that yield poor results.
- A change or turnover in the senior leadership of a business makes it very difficult to establish a lean culture.
- By using lean tools to drive safety improvement, one can demonstrate the power of lean.
- Company cultures, lean or safety, are built on a foundation of trust.
- People do not care how much you know about safety until they know how much you care about their safety.
- Both what leaders say and how they say it is very important.
- The current business processes that established your culture have to change to drive the expected behaviors of the new culture.
- Processes must have owners, and support systems have to be in place to help ensure process stability and improvement.

Chapter 4

Lean Tools for Safety

You Can Continuously Cope or You Can Continuously Improve—The Choice Is Yours

I have heard it said that lean is 75 percent social and 25 percent technical. If that is the case, and I believe it to be true, one's interpersonal skills will be the greatest determinant in the success as a lean champion. The lean tools that will be covered in this and the next chapter (Chapter 5: Advanced Lean Tools for Safety) are not very difficult to understand or use. This chapter will cover some of the entry-level tools used by businesses when they begin a lean journey and how these same tools can be applied to a safety program. In one of the earlier chapters of this book, I noted that many companies initiated lean efforts comprised solely of some of the lean tools and failed in their efforts because they did not address the people or cultural side of their lean implementation. Thinking that lean is about tools is a recipe for failure. Remember, the goal is to engage people in problem-solving activities meant to impact and change the business culture. This type of engagement requires strong social skills, such as coaching, teambuilding, and facilitation. Anyone who uses the lean tools described in this chapter will undoubtedly develop and grow their interpersonal and leadership skill sets.

I have observed many individuals give presentations or conduct tours of their facilities to describe and showcase their lean efforts and I am always amazed at the level of passion they demonstrate. Lean thinkers are by definition passionate people. Yet, they were not born passionate lean thinkers nor did they start their work careers thinking that way. How did the transition occur? What has influenced and changed them? In contrast, it is interesting to note that I rarely see this level of passion from individuals who work in what I would characterize as more standard jobs, such as engineering, purchasing, planning, marketing, accounting, etc. I know many people who work in these professions, some of them passionate, but many are just plodding along. I have often reflected on my perception and wondered what explains this difference in passion for the work people perform. I believe the difference has to do with people, or more specifically, with the growth of people. While engineers, planners, accountants, marketers, and purchasing agents interact with people and follow a standard business process to accomplish their business objectives, lean thinkers go way beyond interacting. They spend their time engaging and growing people. This growth of individuals is a result of engaging them in improvement activities that challenge the status quo of business processes. Those who work in more standard roles see processes as simply road maps of their daily work, while lean thinkers see a business process as the starting point for improvement. As noted in this chapter's heading, lean thinkers have moved beyond continuously coping with (insert your favorite business process frustration) to continuously improving any business process. However, most importantly, they recognize their success is dependent on people. For it is the process experts, the people who work in the business processes, who must be engaged and be part of the change process. If they are not, you have a lean program that will eventually fail.

I was fortunate to learn this valuable lesson concerning working the people side of lean first before getting heavily into the lean tools by working for a company that spent many years, as I described in Chapter 3 building a foundation of trust by engaging all employees in team building. Some of our senior leaders were heavily involved in developing and building our workforce into high performance work teams. Others thought the effort a waste. They could not see beyond the "teams" to the real end goal, which was engaged employees equipped with a laser focus on customer and continuous business improvement. Two successive CEOs strongly believed in developing a learning organization that empowered and engaged the entire workforce. The foundation they built via their leadership and belief in people is still supporting this very successful business today. I believe they understood that businesses are just vehicles to be used to grow people and, if you do a good job of growing people, the business will flourish and the profits will follow. In the end, it is growing people that fuels a leader's passion. For me, my lean passion is a renewable resource. It is my reward for lean leadership. I think that is what differentiates lean thinkers from those who work in more standard jobs. So, how can you harness this combined passion for lean, people and process improvement to build world-class safety? By using the same lean tools and techniques lean leaders use to drive business process improvement.

The engagement tool I am suggesting is "safety." Everyone rallies around safety; hence you will always have willing participants. The unique aspect of safety improvement driven with lean tools is that you also can drive lean improvements at the same time. I know this to be true because I have facilitated multiday kaizen blitz events in which the team's charter, or objective, was to reduce ergonomic injury risks associated with a specific work process. The kaizen teams not only accomplished their goal of making the work processes safer, by reducing both the material handling and physical effort required for the work process, they also greatly reduced the work process cycle time as a result of the same changes. Safety drives lean and lean drives safety. To begin the overview of the lean tools that can be used to drive world-class safety, I will start with the same lean tool most businesses use when they begin their lean journey: 5S (sort, set in order, shine, standardize, sustain).

Lean Tool 5S: A Structured Method of Workplace Organization

I can hear some who are familiar with lean groaning right now. "Please, we do not want to hear any more about this simple approach," they may be thinking. Well, as with all lean tools, the concepts of 5S are indeed simple to understand. It provides a structured step-by-step methodology that will result in a clean, organized work area. A personal experience of mine may help you understand the importance of the visual appearance of a facility. In the early 1980s, when the steel industry was downsizing due to foreign competition and the emergence of minimills, I left the large steel company where I had been employed for the prior 18 years and went job hunting. Eventually, I had two job offers. One was with a government contractor that was maintaining an old ammunition manufacturing facility. The buildings were very old with depressing interiors. The second was a manufacturing facility that was as bright as daylight and very clean. Their core competency was metal stamping. The salary they offered was less than the government contractor had offered, but I accepted the offer anyway. The single most important reason for my decision was that the clean facility seemed inviting and full of promise for a secure future. My decision was due largely on how the facility looked. So, just imagine what potential customers and employees think when they see your facility. Think about perfect organic environments (life plants will flourish with the right light, moisture, and organic materials). Doesn't it make sense that, if you have a well lit, bright, and well-organized workplace, people will flourish? If you think about that long enough, you will understand just one of the benefits of a 5S program.

Many companies have followed the 5S path of sort, set in order, shine, standardize and sustain, and almost as many go back and follow the same path. The reason they may repeat the process more than once is that they do not effectively implement the last two Ss (standardize and sustain), therefore, they have to start over with the first three. Or, to put in into the context of this book, 5S does not become cultural. Instead of implementing 5S because it is part of a lean program or because your work areas are a mess, consider informing your employees you are implementing 5S to improve safety. This is another chance to make "safety first" a reality.

Sort

The first step in 5S is to sort through everything in a work area and identify what is required to do the work, and any related activities, that day. Everything else is questioned, and disposition decisions have to be made for all of it. The disposition options include scrapping, keeping in the area, keeping in another area, selling, returning to the supplier, or donating the items to a charity. It is a renewing exercise for all involved. Compare the sort step to spring-cleaning your garage, basement, or the entire house, if you are really ambitious. Completing any or all of these cleaning tasks gives one the feeling of accomplishment and makes one feel good about the final effect. Take that same approach by telling all your employees you are going to sort through and remove all unnecessary items from the workplace in order to provide a safer working environment. Call it your safety spring-cleaning activity, but follow the 5S lean implementation model. After the sort step, your employees will feel good when they walk into work.

Set in Order

Organizing and identifying all that remains after the sort step is the next 5S activity. The second step is called "set in order" because it requires thought and planning to organize and label tools, inventory, material handling equipment, supplies, paper files, and everything else required to accomplish the work tasks in the work area. From a lean cycle time-gain perspective, this step eliminates the waste associated with looking for things in disorganized workplaces. The benefits are easy to understand in all work environments from offices, with paperwork filing systems, to manufacturing plant floors. To help you understand the safety value of this step, imagine yourself in the middle of the most unorganized manufacturing facility you have ever visited. There is clutter everywhere, no sense or organization, and you are uncertain as to where to walk because no aisle markings exist. It is late at night, you are deep into the plant, and all of a sudden the lights go out. You have no flashlight and have to find your way to safety. Contrast that with a plant that has implemented sort and set in order. The shop floor has minimal equipment, inventory, and supplies. When you arrived you recognized that everything had a place and was located in that place and the aisles were clearly marketed and free of any clutter. Which plant would you like to find your way out of in the dark? In which plant would you feel the safest? How will your employees feel during your next power outage?

Shine

After you have sorted out what isn't required and then organized and labeled all that remains, it is time to thoroughly clean your workspace. You have to go way beyond just sweeping the floor to attain the shine required by a lean focused 5S program. Equipment should be cleaned, painted (if required), and all connections (electrical, hydraulic, and pneumatic) inspected for condition and leaks. I have seen facilities where they paint all equipment white to ensure any leak is immediately visible and repaired. Floors and even walls should be completely cleaned, sealed, and painted. Lighting should be cleaned or replaced. The safety implications are clear. However, the safety implications of this 5S step—shine—go way beyond how everything looks to also include how people feel when they walk into work. I love to hike. Whenever I travel and stay in large metropolitan areas, I like to explore the city. All parts of large cities are not equal. In some sections, usually the areas meant to attract tourists and their tourist dollars, everything is sorted, set in order, and shined. Yet, as I walk and pass through older neighborhoods in these same cities, I come across areas that contain run-down buildings, old cars, unkempt yards, and litter in the streets and yards. In which section of these cities do you think I feel the safest? How do you want your employees to feel when they walk into their workplace? A clean environment sends a strong message that safety is indeed first.

Standardize

This step in the 5S sequence is the opportunity to begin making 5S cultural because this step defines the rules or standards required to maintain the first three steps. Without some kind of a plan for maintenance of the program, as many companies have found, everything will return to its former state. There will be no long-term cultural impact. Different methods of engaging everyone in this step to maintain the 5S gains in their work area can be employed. Written forms, such as check sheets and work area layout drawings, are the most common program documents used to communicate responsibilities. These documents must contain three essential ingredients and, if they do not in 6 to 12 months, you will again be sorting, setting in order, and shining. Those three ingredients are

1. Identify the tasks that have to be completed.
2. Name the individuals who have ownership of each task.
3. Identify the scheduled time during which the tasks will be completed.

Looks like simple task management, doesn't it? I have observed many area or work cell layout drawings where each section of the work area has someone's name listed. This implies that they are to take ownership of 5S for this section of the work area. Figure 4.1a illustrates this oft used, but ineffective, method.

Now, contrast that with Figure 4.1b in which the actual tasks and the time when they are to be accomplished have been added. These two important steps added the work instructions, or standards, to a document that was intended to drive accountability. However, there is still an important ingredient missing. Just because we ask someone to perform a task on a specific day does not mean it will be accomplished. Too often regular work gets in the way. This highlights management's responsibility to ensure time is allotted to perform 5S/safety tasks. It is as simple as designating the last half hour of the day to 5S/safety activities. This standardize step, if implemented correctly, helps to ensure that the workplace safety improvements made in the first three steps have a chance of becoming cultural. The next step can ensure sustainability.

Machine Shop 5-S Plan	
Drills Bruce P.	**Lathes** Gary M.
Mills Henry K.	
	Storage Jerry W.

Figure 4.1a   Standardize plan A: Areas with owners.

Sustain

This is the audit step. Most important business processes have an audit step component. Businesses pay consulting firms for financial audits, which encompass many parts of the business. Included would be inventory control, purchasing, and, of course, accounts payable and receivable. ISO (International Organization of Standardization) audits cover many of the same core business processes looking at the quality of the business systems. Safety audits are conducted to ensure compliance with OSHA and other regulatory agencies. If you expect 5S to truly have a long-term cultural impact, you must conduct 5S lean safety audits. There are countless examples of audit forms available from the Internet and printed lean resources. Many safety programs already have a safety walk component. Any of the lean 5S audit documents can be modified, or combined with your existing safety walk forms, to incorporate safety and then used during regularly scheduled safety walks. They all provide a score so that those in the area being audited get feedback with which to continuously improve. Leadership's role is to conduct or be directly involved in the audits. If they do not accept ownership of this last step, then 5S and the safety improvements it provides will fade away.

Machine Shop 5-S Plan	
Drills Bruce P. 1. Sweep floors daily 2. Empty chip pans every Friday 3. Wipe down machines every Monday 4. Replace missing drills and labels every Tuesday	**Lathes** Gary M. 1. Sweep floors daily 2. Empty chip pans every Friday 3. Wipe down machines every Monday 4. Wipe down chip guards daily
Mills Henry K. 1. Sweep floors daily 2. Empty chip pans every Friday 3. Wipe down machines every Monday 4. Replace missing end mills and labels every Tuesday	**Storage** Jerry W. 1. Sweep storage area daily 2. Inspect and replace any missing or worn labels every Friday

Figure 4.1b   Standardize plan B: Areas with owners, tasks, and frequencies.

Hopefully, this brief overview of the steps of a 5S program has provided the safety reasons for asking everyone to join in this companywide effort. A brief summary is provided in Figure 4.2.

Disciplined Approach Necessary to Maintain Safety in the Workplace

Noteworthy is the strong connection between 5S and safety. Yet too often 5S is seen only as a housekeeping methodology. Those in business, who think that way, fail to understand the greatest value this lean tool provides. The most important element is the creation of a disciplined approach to doing something at work. It goes beyond doing your job. It is additional

Sort	Eliminating clutter, unneeded items, equipment and inventory from the work area reduces the risk of injury from trips and falls and the physical effort required to move this material around
Set in order	Having defined locations for items stored in a work area means you can define clear walkways and workspace
Shine	Clean equipment and floors reduces hazards and makes equipment related safety concerns visible right away
Standardize	The discipline of 5S standards easily translates to the discipline required for a world-class safety program
Sustain	This audit step requirement of 5S can be completed during already scheduled safety walks

Figure 4.2 Safety impact of a 5S program.

responsibility beyond your regular job. It requires thought and action. Compare that to safety. Safety may be part of everyone's job, but it also requires thought. Many accidents happen because people become complacent. They perform the same routine tasks, and do not think, or they are required to perform a nonstandard task and really expose themselves to serious injury because they may not be aware of all of the hazards associated with the task they are about to undertake. There is no better time to have an employee who is a disciplined thinker, who stops and thinks "safety first" before they take any action. 5S can help build that disciplined workforce because it requires people to change and accept change.

It all sounds so simple, yet the acceptance of any type of disciplined change is difficult. Just ask anyone who has tried dieting. A few years ago, I decided the drawer in my kitchen that held the silverware was located in the wrong location relative to the primary point of use. I began by using the 5S methodology to sort the "junk drawer" and the silverware drawer and then swapped their locations. I then applied the 5S theories of set in order and shine to the contents of the silverware drawer. It all seemed to make sense and I was proud of the application of a lean tool in my kitchen. However, the silverware had been in the old location for about 20 years, and that caused a problem. Change is not easy when you have performed a task the same way for a long time. This change provided much laughter in our household for the next three to four months, because we repeatedly, when we wanted a piece of silverware, would go to the old location. We didn't think before we acted. This exercise taught me a lesson about both complacency and how difficult it is for all of us to accept and change our behaviors. Yes, 5S is simple in concept, but it takes a lot of hard work and very committed people to make a plant world-class 5S. I have only been in one world-class 5S plant, but I have toured over 100 plants that were on the lean journey. Because I am a passionate lean thinker, I can talk endlessly about lean and related topics, but when I visited this plant it left me speech-less. There was no need to talk. I could see the application of 5S in every part of the business and, therefore, I was certain it was a safe place to work. This plant was both a sales and marketing tool and a safe place to work. This visit confirmed the importance of building a disciplined workforce. So, use 5S with a focus on safety improvement to build safety discipline. Look at 5S wearing your safety glasses and you will see this lean approach in a new light. It is a great foundation for world-class safety and world-class lean, or lean safety.

Lean Tool: Visual Factory

Anyone who has been involved in the implementation of lean understands the concept and importance of visual methods of communication, which are often referred to as visual factory methods. Visuals make knowledge that previously resided with individuals and in computer systems public knowl-edge. Visual communication methods include signage, charts, signaling cards and containers, lights, and process maps. All of these methods support the lean goal of employee engagement in continual improvement. Let's briefly explore each of these to understand their relevance to lean safety.

Signage

Safety banners with "Safety First" and other slogans emblazoned upon them, safety calendars, safety posters, and other safety visuals have been a long-standing part of almost every safety program. They are part of a market-ing campaign intended to build safety awareness. To me they are part of a "wishing and hoping" safety campaign. They may engage employees when they are first posted, but thereafter they become part of the landscape. Wishing and hoping for a safe work environment does not work. These legacy safety signage methods provide no long-term employee engagement,

ABC Work Cell - Safety Tracking

March						
	Colleen	Sean	Kate	Adam	Liz	Jeff
1	✚	✚	✚	✚	✚	✚
2	✚	✚	✚		✚	✚
3						
4						
6						
7						

Figure 4.3   Cell safety calendar.

which is the key to behavioral-based safety awareness improvement. While benchmarking another manufacturing site, I observed a visual employee engagement safety tool that will work anywhere in the world. An example of this simple but very effective behavior changing visual tool can be seen in Figure 4.3.

Every month, a calendar that contains the names of every employee in a work cell or department is posted in the cell or department. Every day each employee adds a green sticker, with a safety cross printed on it, to the previous day on the calendar if they worked that day without a safety incident or injury. This simple tool requires each employee to take ownership of their safety and report on the results. The action of applying the sticker reinforces a "safety first" mentality in every employee every day. It is a cultural-changing safety process. Compare that with your standard safety calendar that contains a new slogan every month.

Charts

Most safety charting in plants today consists primarily of tracking negatives. Because compliance requirements drive most safety charting, it is injury-related data that is usually collected. Injuries and the cost of injuries is certainly not positive data. I will cover safety metrics in more detail in Chapter 11, but it is important to recognize the need to collect data. Lean thinkers all know you cannot claim improvement if you have not established some baseline metric information. They also understand the reason for collecting process data is to gather the information that can be used to improve a process. Collecting information on the number of injuries and the cost of the same provides no information for improvement. Digging deeper into the same data and tracking the types of injuries can be valuable because it will allow you to focus your safety efforts to prevent those types of injuries. As an example, if back injuries have spiked on the injury-type tracking chart, you may provide lifting training in the departments where this type of injury has occurred. Many current safety programs would take this approach. A lean thinker would go even farther. He or she would schedule a kaizen blitz event to closely analyze the work processes where the injuries have occurred in order to reduce or eliminate the risk completely. Rather than improve lifting, the goal would be to eliminate lifting. The root cause was not the lifting technique, but instead that someone had to lift. Eliminate the lifting and you eliminate the risk of a back injury. Therefore, what you chart is more important than just the act of charting. OSHA regulations will always require some compliance charts. In order to impact your safety culture, you should use visual charts that compile data on proactive metrics that engage your employees in safety improvement.

Signaling Systems

A variety of signaling methods are used in visual lean environments. They range from kanban signals, both cards and containers, which trigger work activity and material replenishment to andon lights that signal the status of a work process via colored lights. Safety systems also have signaling devices. In the Midwest, most facilities have a fire alarm and also a tornado alarm. When sounded, they trigger immediate response no matter if it is a drill or a real event. Lean thinkers attempt to use signals to trigger the same type of response. A flashing andon light that denotes a machine is down and assistance is required should have supervision and maintenance staff responding without being called or told to respond. In the facility where I worked, when a safety situation occurred that required our "first responder team," a paging system called them to the site of the safety situation. I have observed responders drop what they were doing, grab their bag of first aid supplies, and move quickly and directly to that location. Visual signals are valuable tools for lean and safety.

Process Mapping

The last visual factory tool discussed in this chapter is process mapping. This is a powerful lean tool that can be used to engage a team of people in process analysis and improvement. It can be used to visually display the steps of any process so that all can see and understand the process. With understanding comes the ability to contribute to the conversation, and that is engagement. Because any existing safety program is a collection of processes, a skilled facilitator can use the process mapping technique to guide a team or group of individuals through identification and improvement of the safety processes. As an example, a target for improvement could be an accident investigation process. As with all lean process mapping exercises, you would begin by drawing the "current state" map and then proceed to the "future state" map, which would incorporate all of the improvements to the process. Figure 4.4 is a sample process map.

Lean Tool: Training

Yes, you are correct in thinking that training is not just a lean tool but part of every major change that occurs in a business. New software, human resources practices, equipment, and products all require some level of training. Safety training comes in many forms. Some of it, for instance, annual lockout/tagout training, is legislated by OSHA. If you read the following material on the characteristics of adult learners while you recall the experiences of those who participated in the (three-ring binder) kaizen blitz event, you will recognize the value of that learning method.

■ Adults are *autonomous* and *self-directed*. They need to be free to direct themselves. Their teachers must actively involve adult participants in the learning process and serve as facilitators for them. Specifically, they must get participants' perspectives about what topics to cover and let them work on projects that reflect their interests. They should allow the

Completing a Safety Incident Report

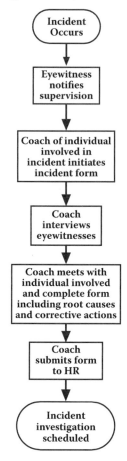

Figure 4.4   Sample process map: Incident report completion.

participants to assume responsibility for presentations and group leadership. They have to be sure to act as facilitators, guiding participants to their own knowledge rather than supplying them with facts. Finally, they must show participants how the class will help them reach their goals (e.g., via a personal goals sheet).

■ Adults have accumulated a foundation of *life experiences* and *knowledge* that may include work-related activities, family responsibilities, and previous education. They need to connect learning to this knowledge/ experience base. To help them do so, they should draw out participants' experience and knowledge, which is relevant to the topic. They must relate theories and concepts to the participants and recognize the value of experience in learning.

- Adults are *relevancy-oriented*. They must see a reason for learning something. Learning has to be applicable to their work or other responsibilities to be of value to them. Therefore, instructors must identify objectives for adult participants before the course begins. This means, also, that theories and concepts must be related to a setting familiar to participants. Letting participants choose projects that reflect their own interests can fulfill this need.
- Adults are *practical*, focusing on the aspects of a lesson most useful to them in their work. They may not be interested in knowledge for its own sake. Instructors must tell participants explicitly how the lesson will be useful to them on the job.[1]

Doesn't it seem like the author of this material is describing the teaching methods and learning opportunities presented during a kaizen blitz event? Because kaizen events are hands-on learning and doing events, safety professionals who deliver safety training regularly see the value in this lean approach. The loss control representative and a professional ergonomist from the member trust I referred to earlier, who have been involved in the safety kaizen blitz events that I have facilitated, have concurred that the opportunity to almost immediately apply the lessons taught by going to the shop floor and focusing on safety improvement is priceless. They noted that their past experiences usually followed this script. They deliver the safety training on-site and then leave. When they return for their next visit, nothing has changed. Despite the good intentions of those who received the training, the excuse most often heard is: "We didn't have time." By mirroring the lean method of training, which is almost always a bit of classroom mixed with a large dose of hands-on experience, you can fulfill the needs of adult learners and immediately start to see the results of your training.

Lean Tool: Poka-Yoke (Failsafe)

No, this is not a dance performed at wedding receptions (that is the hokey-pokey). It is another Japanese-titled lean tool, which roughly translates to *failsafe*. The application of this lean tool normally focuses on methods that ensure the quality of manufactured parts. If you can devise a method that will make it impossible to perform an operation incorrectly, then you have identified a poka-yoke. For instance, a simple dowel pin inserted somewhere within a part that holds a fixture, which would allow the part to be inserted in only the correct direction would be a poka-yoke. Can this method of ensuring quality be applied to safety? Is it possible to be 100 percent certain that someone cannot be injured when performing a task? If you read the English translation of the term *failsafe* and take the word literally you might be tempted to immediately answer yes. I am not that sure you can answer yes because humans can and do disarm or disable safety devices that are considered failsafe. Light curtains, bolted on guards, and interlocked entry doors are all intended to make it impossible for someone to get their hands into the danger zone when a machine is cycled. There is an entire industry working on developing, marketing, and selling machine guards. Yet guards are removed, interlocked entry door switches are blocked with pieces of tape, and light curtains bypassed. A safety device is only as good as the safety awareness of the operator. They have to work in harmony to prevent serious injury. Just as a company staffed with nothing but lean thinkers is a goal, having a company where everyone, at all times, thinks safety before taking any action is also a goal. To move toward that safety goal requires changing the approaches currently taken. As we explore current methods employed to raise safety awareness in future chapters and then contrast them with approaches influenced by lean thinking, you will begin to understand how to move your safety program closer to that end goal.

Lean Tool: Benchmarking

Benchmarking has been a practice long before anyone uttered the word "lean." The formal application of this method of learning and sharing learning with others has most often been utilized by larger corporations. An everyday role of a friend who was employed by a large aerospace firm was to lead benchmarking efforts for the corporation. Benchmarking was a word used in his job title. He and others in his firm developed processes and forms that could be used to manage the benchmarking process within their business. He co-authored a book about "knowledge transfer," which is what benchmarking is intended to accomplish. My guess is they used that title to differentiate their book because you will find many pages of books with benchmarking in their title when you search Amazon.com. It is common sense to learn from others. Most companies, however, do not establish formal benchmarking programs, but instead look for other opportunities to learn.

Businesses or individuals who are on the lean journey exploit plant tours and lean conference presentations to harvest new ideas they can take back and apply in their plants. "Steal shamelessly" is a term that I believe may have been coined by the lean community and they take it to heart. Plant tours that are often referred to as "industrial tourism" are a great opportunity to benchmark what others are doing. They provide the chance to see as well as hear about other's lean efforts. Often those on plant tours walk through a plant with only an open mind and eyes and collect what they can. Benchmarking is not like that at all in that specific objects are defined before another facility is visited. It is industrial tourism with a target. So why not target safety? The company I worked for utilized the fact that it was a member of an insurance trust that contained around 100 companies to their advantage. The trust offered safety-training programs, held quarterly meetings, often at a member company site, and an annual meeting. At all of these events were individuals responsible for or involved in safety at their plants. Often, as a result of the contacts made at these events, arrangements were made for informal benchmarking visits to other trust member sites. Effective networking at any of these events is just like having access to free consulting. The trust was our company's safety community and we used it to our advantage.

Lean Tool: Continuous Flow/Cycle Time Gains

Because the goal of lean is to reduce the cycle time between paying and getting paid, whenever processes are linked, the total cycle time is reduced. From a safety perspective, whenever you link processes, you reduce the requirement to move material. Less material handling equals a reduced safety injury risk. Whenever lean is implemented, safety is positively impacted because the drive to reduce cycle times often makes the work performed easier for the operator. Easier generally means safer.

The point that reducing cycle times yields easier ways to perform the work observed was better understood after two kaizen events held at one of our company's distributor's locations. The focus of both of the events held at their site was cycle time reduction as opposed to safety improvement. The first kaizen team's charter charged the team with reducing the cycle time to process a length of industrial belting. A lot of physical effort was involved in moving and prepping the belting. As the kaizen team observed one of the steps, stripping off strips of the top and bottom cover of the belting, we all immediately recognized the requirement to reduce the physical effort. Belts are composed of a fabric core (the carcass), which provides the strength of the belt that is then covered with rubber or some plastic compound. To prepare these belts to the customer specifications two-inch-wide strips of the top cover and the bottom cover had to be pulled off the belt carcass at the cut ends. To begin the process, the belt cover was cut down to the carcass using a steel ruler to guide a razor knife. Once the strip of belting to be removed had been scored across the width of the belt, an air chisel with a flat blade tool was used to chisel up one end of the belt cover strip. Next the belt was thrown onto the floor. Now, the belt technician, using a pair of gripping pliers and standing on the belt to hold it down, latched the pliers onto the flap that had been chiseled up and, using his back, pulled up to release the strip of belt covering from the carcass. He had to repeat this step multiple times because he was pulling on a large rubber band that stretched as he pulled. When he completed one side of the belt, he had to flip it over and repeat this operation on the bottom side. I noticed that the belt technicians had tremendous upper body strength and now I understood why. Our goal was cycle time gain, but in my mind, we had to find an easier and, therefore, safer way to remove the belt cover. In this case, the team turned to technology to solve the problem. They developed a custom tool that fit into a ½-inch portable drill that had high torque and a slow speed. The belt now stayed on the bench where it had been scored and the belt end was clamped to the table. The flap of belting that had been chiseled up was inserted into a slot in the custom tool. As the drill rotated the tool, the belting was peeled off of the carcass and rolled onto the tool. If you visualize the old sardine cans with the key that was used to wind off the can cover you will understand how this worked. Kaizen blitz events are often criticized because some changes do not stick; people go back to the old way. Not so in this case. When you implement real changes that genuinely make someone's job easier, they are not going back. This was especially true in this case because the belt technician had not been a big fan of six people watching and telling him how to work. After this new process was tested and proved, he "came over" to the lean side. The goal had been cycle time gains, and they were achieved, but in my mind the biggest reason for these gains were this and other safety improvements made during the event. Let's review the second case study from this same site.

The charter spelled out a goal for this second event that charged the team with reducing the changeover time on a large piece of equipment. On the first morning, the team observed the current state changeover and recorded the cycle times. Total cycle time amounted to around 80 minutes. One of the tasks performed during the changeover had been to climb up into the center of the machine, between and the upper and lower sections, to change shim sets that controlled the stroke of the machine. This machine was hot since it was used to cure rubber belting. Working in the center of the machine was akin to working in a car that had been sitting in a hot parking lot in August in Miami. The machine operator worked fast and efficiently and yet it had taken him around 20 minutes to change six shim sets. This involved lifting six individual steel blocks that weighted around 40 pounds each to remove and then replace shim sets that sat underneath each of them. The solution the team arrived at was brilliantly simple. The shim sets were placed on top of the steel block so the block no longer had to be removed. During the future state changeover that was observed on the last afternoon of the event, this step, or to say it another way, the operator's time in the hot car, had been reduced to 1 minute and 20 seconds. No more lifting of heavy objects either. This operator is not going back to the old way.

In both case studies, the team's lean goals to reduction cycle times were met. In both cases, the cycle time gains would not have been possible without the great safety improvements the team identified and implemented. After the team presentation to management in which the second case study results were reviewed, the president of this company asked to make a few comments. He first congratulated the team on the cycle time gains it had made (he had an accounting background and was already calculating the machine capacity gain the event had yielded), and then he said he was amazed at how this event had positively impacted safety. Now there is a leader who understands lean safety. Lean and safety are as tight as a carcass inside of a belt cover.

All of these examples support the fact that lean implementations drive safety improvement even when safety is not the primary objective. And, as described earlier in this chapter, safety improvement efforts that are undertaken using lean tools, like the kaizen blitz, drive lean or cycle time gains.

Lean Tool: Standard Work

Unless a company gets to the point where everyone from leadership to the shop understands the essence and importance of standard work, lean will not become cultural. Standard work defines how the work will be performed. Documents that describe the work methods and expected outputs are the building blocks used to establish a culture where standards are clear, employees feel productive, and consistent results expected. Standard work is the basis for establishing a continuous improvement culture where everyone is engaged in continuous improvement. Recently I conducted some lean training I called "Lean 101." One of the exercises the participants were asked to complete was to write the definition for an "engaged employee." Each table of attendees was allowed to work collectively on the definition. After about five minutes, I asked each table to report. They used terms, such as *involved, having a good day, working with others*, etc., in their definitions. As I probed to get to a deeper meaning, I asked them if they could describe their "best day at work." I repeated this class many times and invariably the general response went something like this. "Everything went really well, the equipment ran all day, and I felt productive." I then would ask them if they could tell me the production rates, e.g., pieces per hour, which they were expected to meet. Most guessed, but did not actually know. They could not tell me the standard that should define a good or a bad day at work. How could they possibly know if they had a good day if they had no metric to define "good"? This is a management problem because management had not clearly defined standard work for their employees and without standards there is no basis for continuous improvement. Contrast that with a distribution center where the order pickers know that standard work requires them to make 12 picks every 15 minutes. In any of those 15-minute blocks of time when an order picker cannot meet standard, they throw a switch that turns on a signal light to alert their supervisor. The supervisor responds immediately—not to focus on the person, but to focus on the process with the order picker. Together they immediately begin problem solving. Was a part misplaced, was the quantity on the shelf incorrect, were the parts stored incorrectly? This is an example of a continuous improvement culture where everyone understands the standards and they engage in continuous improvement to maintain the standards. In the first scenario, the employees come to work, do their best, and then go home feeling unfulfilled. If managers are not honest and definitive about the standards, they do their employees a disservice. They have denied a basic need everyone has—to know if they added value based on some standard. This is not just a problem related to production rates. It is a systemic problem that goes deeper. Many managers and especially supervisors shy away from giving honest feedback to their reports on almost any topic. They feel as if it may lead to hard feelings that will damage their relationship with the employee. So, instead, they tolerate poor performers for long periods of time. But, what they should understand is that they are denying their reports the opportunity to grow by not providing honest feedback. They are missing the opportunity to coach their reports to a higher level. Standards must be defined, communicated, and enforced, not to police people, but as a means and basis of collaborative continuous improvement.

Are there or should there be standards when it comes to safety? Absolutely. Lockout/tagout training clearly defines standard methods employed to isolate stored energy before someone begins to work on a piece of equipment. As a matter of fact, OSHA spells out hundreds if not thousands of safety standards for many industries that are best practices. Some are very specific and many are general in nature, but they are all intended to prevent workplace injury. These are mandated standards and can be considered standard work. To be compliant with these regulations requires knowledgeable staff, training for all employees, and ongoing vigilance in the form of audits. OSHA compliance activities, though necessary, do not engage everyone in continual safety improvement.

There is a program, however, that will accomplish that cultural objective. It is titled Job Safety Analysis or simply JSA. A JSA is a method that can be used to identify, analyze, and record the steps involved in performing a specific job, the existing or potential safety and health hazards associated with each step, and the recommended actions and procedures that will eliminate or reduce these hazards and the risk of a workplace injury or illness. Think of this safety program as a series of mini safety kaizen events. As traditionally used, a JSA program is a pure safety activity and, in plants where they have decided not to engage their workforce, they use their engineers to develop the JSAs. When the documents are completed they are great training tools useful in training anyone new to the work tasks. The finished JSAs are "standard work" documents that cover only the safety aspects of the work process. In a lean culture, establishing a JSA program would be a great vehicle to engage everyone in safety improvement. If management decides to pursue that objective, a process owner must be assigned and support systems put in place to ensure the program's success. I was part of a JSA program startup that dwindled and died because it had no process owner. It was another case of distributed ownership of a new program. That always ensures failure. I have attached a simple version of a JSA form (Figure 4.5) for review. More detailed versions are available online at http://www.ccohs.ca/oshanswers/hsprograms/job-haz.html.

Lean Tool: Problem Solving

"Asking 'why' five times" is a common problem-solving methodology used in the lean community. It is a way to get to the root cause of a problem. Be forewarned that it can also result in heightened emotions. People are not

	Job Safety Analysis				
Equipment No./Name		Department Name			
JSA No.	Date Written	JSA Author(s)			
PPE required for this JSA -					
Step No.	Description of Job Step	Hazards Present	Action to Prevent Injury	Training Tips	

Figure 4.5   Job Safety Analysis (JSA) sample form.

accustomed to having someone grill them by asking "why" over and over. I often inform the person about the technique before I press them a bit to get to the root cause. Five is not a magic number; the number of times you ask the question is situational. It is very effective at framing the conversation and engaging all involved in root cause analysis. Therefore, it is a very useful technique when conducting incident or accident investigations. Chapter 7 will be devoted to that important aspect of world-class safety.

Another problem-solving methodology popular with lean thinkers is the PDCA problem-solving loop. PDCA (plan-do-check-act) is an iterative four-step problem-solving process typically used in business process improvement. It is also known as the Deming cycle, Shewhart cycle, Deming wheel, or Plan-Do-Study-Act.

1. *Plan*: Establish the objectives and processes necessary to deliver results in accordance with the expected output. By making the expected output the focus, it differs from what would be otherwise in that the completeness and accuracy of the specifications is also part of the improvement.
2. *Do*: Implement the new processes.
3. *Check*: Measure the new processes and compare the results against the expected results to ascertain any differences.
4. *Act*: Analyze the differences to determine their cause. Each will be part of either one or more of the PDCA steps. Determine where to apply changes that will include improvement. When a pass through these four steps does not result in the need to improve, refine the scope to which PDCA is applied until there is a plan that involves improvement.[2]

The value of this tool is that you can continue to repeat it over and over again on the same process. Here is an example based on my experience of how it can be used within a safety program. An end of the year review of all safety incidents revealed that the category with the highest number of incidents was forklifts incidents. Follow the PDCA actions below that were used to turn this safety concern around.

■ *Plan*: Increase the amount of training for forklift operators by bringing in an outside training firm. The goal is to reduce the number of forklift incidents.
■ *Do*: Training is conducted and all forklift drivers attend.
■ *Check*: The incident log is monitored and by September no appreciable difference in incident rates has occurred.

- *Act*: The plant general manager decides further actions are required to reduce the forklift incident rate.
- *Plan*: Plant general manager decides that he will lead an incident investigation for all forklift incidents in order to get to root causes and corrective actions.
- *Do*: The plant general manager leads the investigation of all forklift incidents for the next 10 months.
- *Check*: Forklift incident rates have dropped by 50 percent.
- *Act*: Continue with the investigations to drive the incident rate lower.

This continuous improvement tool, a never-ending improvement loop, can be part of your safety culture. It became part of ours because anyone could simply say, "This process needs to be PDCA'd" and it was understood by all.

Lean Tool: Metrics

Measurements or metrics are part of every business. Most are financial because the success of a business is measured in dollars and cents or some other currency (I am hoping for some international sales). A lot of continuous improvement or lean activity is touchy-feely and requires a certain amount of faith that what you are doing will result in good things down the road. Some leaders understand this, but most do not. Our business schools train MBAs to focus on dollars so, when it comes to programs that espouse teamwork and empowered employees, they don't get the relevance of these activities to the balance sheet. So, like it or not, lean champions and senior level lean leaders must measure and prove results for the efforts they lead. Safety programs, on the other hand, have always had a metric component. Most of the metrics are compliance driven. Below is a brief overview of the OSHA recordkeeping guidelines.

- The 300 log is a listing of all the injuries and illnesses at your site.
- The 301 form is the individual record of a work-related injury or illness.
- Form 300A is the summary of work-related injuries and illnesses. This is the one you post every year.

Because there are penalties for noncompliance, companies do keep their OSHA-required metrics current, but many stop there. In Chapter 11 on metrics, I will explain how using metrics that are related to innovative safety awareness programs can measure as well as promote ongoing safety improvement.

Lean Tool: Teams

In the late 1980s and early 1990s, teams and teamwork were at their zenith. It was one of the foundation pillars of world-class or continuous improvement programs. Companies were jumping on the team bandwagon like they have boarded the lean train recently. The wheels usually fell off the wagon after a short time because management discovered a big problem—implementing high-performance work teams meant they had to change the culture of their businesses. This was a multiyear task and most did not have the patience or passion to see it through. That is when the continuous improvement community (the word *lean* had not yet been coined) started using some of the TPS (Toyota production system) tools, such as 5S, SMED (single minute exchange of die), TPM (total production maintenance), etc., to drive and champion change, and teamwork fell out of vogue. Many companies worked very hard using these tools to impact their business processes long term. Most failed. They failed, but then realized that these changes were not going to stick unless they became cultural. It doesn't matter if you go by wagon or by train, you have to change the culture of your business to reap long-term benefits. Let me explain this all-too-often condition with the following example.

A 5S program defined by management and a consulting group is pushed onto the shop floor. Management has delegated responsibility for the program to the consulting group and their internal lean champion. The consultants provide training for everyone and then a pilot cell is 5S'd. Management is invited to a presentation followed by a trip to the shop floor, which makes them uncomfortable because they rarely visit, to see the results of their 5S program. They are impressed, thank everyone for their efforts, pat themselves on the backs, and get back to the important business of running the business. The consultants leave and the internal lean champion is expected to expand the 5S program throughout the plant. He gets nowhere because he has no management support. Management calls him in and fires him because he has been unsuccessful at leading the 5S program implementation. This is why lean champions were changing jobs frequently in the 1990s. Then, after about 15 years of moving from TPS tool to TPS tool, the lean community came to a consensus that for change to stick, you have to impact the culture. With that realization, business leaders that were serious about leading change in their businesses accepted the fact that their direct involvement was a necessity, not an option.

In Chapter 3, I mentioned teamwork being a significant element of the continuous improvement effort where I was employed. The two senior leaders of the business at that time understood that teamwork was not just a fad and that it could be used as a cultural change tool. The teams were not the end goal—teamwork was the engagement method selected to drive culture change. This cultural change had to take the business from one with an uninvolved "entitled" workforce, a directive supervisory style, an office versus factory culture, MRP planning, and a lack of universal customer focus caused by a silo organizational structure to something radically new. Leadership also understood very clearly that they had to be directly involved for a cultural shift to occur. In one of the most visible demonstrations of leadership I have ever witnessed, the COO of the business moved from his cushy office space on the second floor with a view of the nicely landscaped property surrounding the business, to a desk that was surrounded by yellow pipe barriers for protection from the forklifts that passed routinely, and situated in the middle of the shipping department. We had just embarked on a journey to empower and engage all of our employees and the vehicle chosen for the journey was high-performance work teams. Our COO understood that, for such a radical change to be successful, he had to lead the effort. He became the coach of the first high performance team: our shipping team. During his tenure as their coach, he was promoted to CEO and yet he stayed put. He stayed put until he passed the company leadership role to someone else and moved on to pursue other passions. His leadership, combined with the dedicated work of many other people, transformed the business culture. Teams were not the objective, just the vehicle to establish an empowered workforce, which staffed customer-focused work cells and used pull signals and time relevant metrics to ensure customer satisfaction. This cultural shift connected the workforce to something bigger. They could no longer just come to work and check their brain at the door, repeat the same work tasks they performed every day, and then pick it up when they departed. It both helped and forced people to change, by getting them involved in many aspects of their team-building efforts and company business processes normally outside of the scope of their work. Everyone grew through this effort and some that did not fit into this new culture left. Some results of this multiyear effort were customer service and productivity metrics that both spiked to world-class levels. This business had developed, via the hard work of many people, a world-class culture when most businesses were still experimenting with lean tools. Improvement efforts, no matter if the focus is lean or safety, fail unless there is direct involvement from senior leadership. Sailboats do not steer themselves nor do business change efforts.

Teamwork is an exceptional lean tool. It can quickly provide a common focus for a group of people. Chapter 5 (on advanced lean tools) will cover a best practice example of using safety-focused kaizen blitz teams. Using a team structure as a foundation pillar to build world-class safety is a sound approach.

Quick Guide: Lean Tools for Safety

- Lean is 75 percent social and 25 percent technical.
- Engaging employees in continual improvement requires strong social skills, such as coaching, teambuilding, and facilitation.
- Lean champions spend their time engaging and growing people because they recognize their success is dependent on people.
- Use safety as a people engagement tool to drive business culture change.
- Lean tool application in safety programs:
  - 5S: Use to build a safety first culture. Creates a discipline that is the foundation for lean. Turns a plant into a sales and marketing tool.
  - Visual factory: Use safety visuals to engage employee base in "daily" safety activities.
  - Training: Lean training, like the hands-on participation during a kaizen team, is the best form of training because you are immediately applying what you are learning.
  - Poka-yoke: Failsafe methods combined with heightened safety awareness can help to ensure an injury-free workplace.
  - Benchmarking: Steal shamelessly from others to improve your safety program.
  - Continuous flow: This lean tool helps to eliminate the material handling and material movement so common in batch systems.
  - Standard work: This is the foundation piece of continual improvement, safety, and respect for people. A focus on mandated safety standards, like JSAs, can help people understand the "standard work" lean concept.
  - Problem solving: Asking "why" five times can be used in accident investigations to get to root causes. A PDCA problem-solving loop can be used to redefine safety processes.
  - Metrics: Defining proactive safety improvement programs and metrics will drive you toward world-class safety.
  - Teams: Build a safety team to engage your employees in safety improvement.

Endnotes

1. Lieb, S. (1991) Principles of adult learning. *VISION*, Fall: http://www.hcc. hawaii.edu/intranet/committees/FacDevCom/guidebk/teachtip/adults-2.htm
2. Wikipedia (2009) PDCA (define): http://en.wikipedia.org/wiki/PDCA_cycle

Chapter 5

Advanced Lean Tools for Safety

Facilitated Leadership: Changing How People Think

By combining some of the basic tools covered in the last chapter with an improved awareness of facilitated leadership, you can start using some advanced lean tools to engage teams of employees in safety improvement. Two will be described: the A3 and the kaizen blitz.

Lean Tool: A3 Problem Solving

The A3 is both a lean leadership development and an employee engagement tool. This problem-solving project management methodology merges activities, such as process mapping, problem identification and solving, goal setting, and auditing, and does it all on one piece of paper, front and back. It is the perfect tool for managers to use because it gives them a structured approach that can be used to guide a team to their outcome. "Their outcome" implies that the leader of the A3 exercise must play the role of a facilitator. Lean leaders understand that managers are best when they manage processes and lead people. As noted earlier, the emotional intelligence, or interpersonal skills, of the facilitator is a critical element required to use this tool effectively. Guiding the team by having the right questions, not the right answers, is difficult to do if you have historically managed your people and are not familiar with facilitation techniques. The real value of this tool is that it will help anyone become a better facilitator if they just follow the six-step structured approach it offers. Figure 5.1 is a sample of a blank A3 that you can review as I describe the sequential steps that follow.

Step 1: State the Agreed Upon Problem or Need

The key for the manager/facilitator in this step is to not offer up the problem statement before the team has a chance to discuss and agree upon the problem statement. If a manager defines the problem to be solved up front, then the team is likely to just go along. He/she has lost the opportunity to engage this group of employees in real problem solving. They are now just passengers on the A3 bus. The key is to carefully craft questions to engage them so that they reach a consensus. If they see it as their problem statement, they will be engaged in solving the problem. Remember, a facilitator is much like a baseball pitching coach, they assist but they do not deliver. It is also important to make certain the problem statement has a customer focus. Since you are using the A3 to assess and improve a business process, recognize the customer, internal or external, of that process in your statement. Another best practice to guarantee you have a good problem statement is to review it using the SMART test. Is it specific, measurable, attainable, relevant, and time bound? Once the team has agreed upon their problem statement, it is time to take them to step 2.

Step 2: Draw the Current State Map or State the Current Condition

Step 2 is another opportunity to hone facilitation skills. Using simple process mapping symbols to spell out the current state process, the team has the opportunity to interact and identify problems and opportunities for process improvement as they complete the mapping. The fact that process mapping is visual helps to engage everyone because the map is drawn in full view of all. That is one of the real values of the A3. On one piece of paper, the entire project is laid out in front of everyone.

Step 3: Problem Solving—Use Ask "Why" Five Times Technique to Get to Root Causes

As the team mapped the current state in step 2, they built a list of problems and opportunities for improving the process. Utilizing that list and the

A3 Report

Step 1 – State agreed upon problem or need

Name: _____

Date: _____

Step 2 – Current state – Process map or list current condition

Step 3 – Problem solving – Use ask "Why" five times technique to get to the root causes. List problems or identified opportunities

Figure 5.1 Blank A3 form.

Step 5 – Implementation plan

	Task	Owner	Due date
1			
2			
3			
4			
5			
6			
7			
8			
9			
10			
11			
12			
13			
14			
15			

Step 6 – Audit expected results against actual results

Auditor	Date	Results

Step 4 – Future state – Draw a process map or describe how you would like it to be in the future

Figure 5.1 (continued).

recently completed map, the facilitator uses questioning, such as the often-used "ask 'why' five times" technique, to guide the team to the real root causes of the process problems. As the questions and resulting conversations take place, a refined process, based on the clarified opportunities identified, will start to take shape. It is now time to take the team to step 4.

Step 4: The Future State—Draw a Map or Describe How You Would Like It to Be

Using the list of opportunities defined in step 3, engage the team in drawing the future state map. The facilitator's role is to guide the team to a future state that is both clear and can be accomplished. This is their vision for how it can be and by completing it together, as a team, buy-in results. It is critical for this ownership to occur because the team will be charged with completing the tasks required to get to the future state.

Step 5: Implementation Plan

Now, build a task list to include all of the actions required to get to the future state condition. Each task should have a defined owner and a target date. The value of this step is the accountability and task management it provides. With the future state clearly in mind, the team can be turned loose to accomplish the identified tasks. When all tasks have been accomplished and the future state process has been in use for some period of time, the final step of the A3 process can be completed.

Step 6: Audit Expected Results against Actual Results

Referring to the task list from step 5 and the future state map from step 4, an audit should be conducted to confirm task completion and that the future state process steps are in place and functioning as expected. After a successful audit, call the team together to celebrate their success. Build contagious commitment by always celebrating these successful events that help to build the new culture. Together view the completed product and historical record of their accomplishment—the A3.

There are books written about the A3, its origins at Toyota, and its power to transform leaders and businesses. I have visited and toured two facilities, one manufacturing and the other medical, that have chosen the A3 as their standard method of problem solving. The A3 is used for all continuous improvement projects, small or large, at these facilities and the results were impressive. The A3 was part of the culture in both facilities. Here is an example of the depth of the cultural reach of the A3. Those of us attending the A3 workshop, at the medical facility, were given a facility tour that included stops at which individuals and teams presented their A3 projects. At one of our first stops, the housekeeping staff talked about how they used the A3 to make certain they always had working vacuum cleaners. At the last stop, a member of the hospital's board of directors displayed the A3 he used to engage the department heads in developing standard board reports. The event host made it very clear that at this hospital the A3 was part of the culture from housekeeping to the boardroom.

My goal was to provide a broad overview of the A3 structure, its use as a simplified method of project management, and to suggest it as a method to improve safety processes. The A3 has the power to transform you into a skilled facilitator and transform any safety process you select. Figure 5.2 is a completed A3 chart to help one better understand this lean tool and its uses.

Lean Tool: Kaizen Blitz

The kaizen blitz, or a rapid continuous improvement event, is the most powerful people development and engagement tool in a lean thinker's toolbox. These events have the power to transform a business culture. In Chapter 3, I described one of my first experiences at facilitation of a multiday kaizen event and how that experience changed me. Some of my most fundamental beliefs, a belief in people, a belief in focusing on process, and a belief that continuous improvement is indeed endless and fun, were all confirmed during that one event. If you haven't experienced the almost magical effects on the participants in a successful kaizen event, you have missed the most fun there is in manufacturing. These events provide the renewable fuel for my lean engine. I will walk you through the entire process of using this powerful tool to engage people in business improvement.

The use of kaizen blitz events is nothing new. Kaizen blitz events were being held before the term *lean* (in this context) became part of our vocabulary. New buzzwords, for continual improvement activity, e.g., lean, Six Sigma, and lean sigma are continually being created by authors and consultants in order to sell books and services. (I am guilty as charged for creating lean safety.) So, it is not the name of a company's program that is important. Many companies, because they do not want to use someone else's name for the continual improvement of their business, create their own name, such

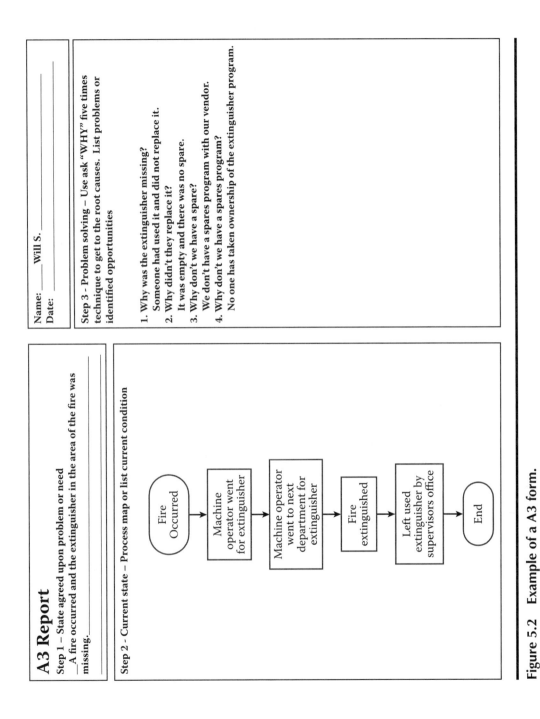

A3 Report

Step 1 – State agreed upon problem or need
  A fire occurred and the extinguisher in the area of the fire was missing.

Step 2 - Current state – Process map or list current condition

Fire Occurred

→

Machine operator went for extinguisher

→

Machine operator went to next department for extinguisher

→

Fire extinguished

→

Left used extinguisher by supervisors office

→

End

Name: _____ Will S. _____
Date: _____

Step 3 - Problem solving – Use ask "WHY" five times technique to get to the root causes. List problems or identified opportunities

1. Why was the extinguisher missing?
   Someone had used it and did not replace it.
2. Why didn't they replace it?
   It was empty and there was no spare.
3. Why don't we have a spare?
   We don't have a spares program with our vendor.
4. Why don't we have a spares program?
   No one has taken ownership of the extinguisher program.

Figure 5.2 Example of a A3 form.

Step 5 – Implementation plan

	Task	Owner	Due date
1	Set up extinguisher spares program	DJ	6/25/10
2	Define spares location in Stores	VK	6/25/10
3	Train all employees in new process	KH	6/30/10
4	Fire drill with extinguisher usage	AJ	7/15/10
5			
6			
7			
8			
9			
10			
11			
12			
13			
14			
15			

Step 6 – Audit expected results against actual results

Auditor	Date	Results
RBG	7/15	During fire drill extinguisher process was observed and worked as planned.

Step 4 - Future state – Draw a process map or describe how you would like it to be in the future

```
Fire
Occurred
   ↓
Machine
operator went
for extinguisher
   ↓
Machine
operator
extinguished fire
   ↓
Used extinguisher
taken to Central
Stores and exchanged
for a full one
   ↓
Full extinguisher
put in correct
department
location
   ↓
Central Stores
sends used
extinguisher out
for recharge
   ↓
End
```

Figure 5.2 (continued).

as the ABC Company Production System. Thus, the program titles may vary, and the lean tools like the kaizen blitz, may be referred to as "rapid continuous improvement" events, but the intent is the same. All businesses that engage in continual improvement want just that and kaizen blitz events are a proven method that I believe can jump-start or reignite lean in any business.

I know of companies who hold kaizen blitz events monthly or even weekly to drive change throughout their entire facility. There is at least one book whose subject matter is entirely about the kaizen blitz and the benefits of using it. The approach is generally always the same, but the kaizen events may vary in length depending on the process being evaluated and improved. They are always team-based activities with a cycle time reduction focus. This focus on cycle time supports "lean" as defined in the definition I most often use. "Lean is a manufacturing philosophy that shortens the time line (total cycle time) between taking a customer order and the delivery by eliminating waste." This widely used and accepted method for people engagement and business process improvement does not need to be redefined or structurally changed. What has excited me, that I wanted to share in this book, is a use for this lean tool that will help to build world-class safety. How? By forgetting the focus on cycle time during the next, or maybe the first, kaizen blitz event and instead making the primary focus safety improvements. During these lean safety kaizen events, no stopwatch is used, thus, helping to ensure that the focus is on safety and not on cycle time. What inspired me to write this book were the results of lean safety kaizen blitz events that I have facilitated. Those experiences taught me that the true path to lean is through safety because safety is and always should be first.

Planning the Kaizen Event

Figure 5.3, which is a process map of the planning steps required prior to beginning a kaizen blitz event, can be referenced as I describe the steps in more detail. It is often someone else, an event champion or sponsor, who suggests the targeted process to be improved via a kaizen event. Therefore, the first and very important step is to evaluate the process suggested for the event. This evaluation can only be accomplished by going to the process and directly observing what is to be improved. Facilitators will want some specific questions answered as they conduct this initial process analysis. If they are planning a lean- or cycle time-focused kaizen, they will want to ensure they understand the process steps involved in the current state

Kaizen Blitz Step 1 - Planning the Event

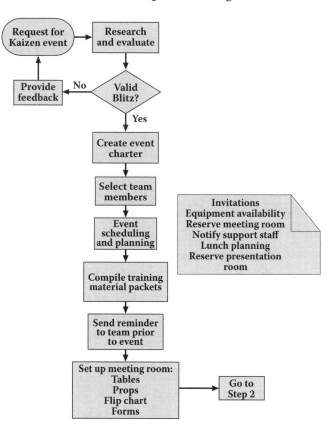

Figure 5.3   Kaizen blitz map 1.

condition. They will also, based on their lean experience, be looking for any "waste," or potential improvements, evident in the process. A facilitator wants to ensure the process is robust enough to challenge and engage a team for three days. These two observations are also important because they will help later when together the facilitator and event champion select the people to staff the team and determine its size. Once consensus has been reached with the event champion that the targeted process is right, the facilitator's attention can be directed to creating the team's charter. However, before moving to that next step, what would be different if the event champion's request had been to facilitate a lean safe kaizen blitz? How does planning an event where the focus will be safety rather than cycle time change this step? The primary difference is in the assessment of the current state process steps. As these work steps are observed, the focus would be to identify safety improvement opportunities primarily related to ergonomic

injury risk reduction. Therefore, every material handling or assembly operation observed would be a potential area for safety improvement. If the process observed contains many of these steps, the facilitator would be assured
that the team will have adequate opportunity to make a safety impact during their kaizen event. So, whether the event's focus will be cycle time or
safety, a team charter is the required next step. The team charter (Figure 5.4)
defines some very important guidelines for the team.

Starting at the top is the name of the process they will observe and
improve followed by an agreed upon starting and ending point of that process.

It is important for the facilitator to keep the team within some boundaries, or the scope of their assignment together. When they start discussing
and pursuing improvement opportunities outside of their scope it is called
scope creep.

If that occurs, the team may be biting off more than they can accomplish
in the time allotted. The facilitator will use these agreed upon guidelines
to rein the team in when they start to stray. The next line of the charter
requires a goal for improvement to be set. The event champion and the
facilitator usually agree to this goal before the team is formed. Goals should
not be easily attained. They should encourage the team while at the same
time stretching their current thinking so as to constantly encourage them
to go one step farther. It is natural for a lean cycle time-focused event to
select a percentage-type improvement. For example, if a team was charged
with reducing the cycle time of a multistep assembly operation, or an office
team's goal was to reduce the customer order paperwork processing time, a
50 percent improvement target for either team would be appropriate. Targets
of 50 percent or more are often the norm during kaizen events. How does
this metric-based, goal setting translate for a lean safety event? Easily. Just
target a number of safety improvements that will reduce ergonomic risks
in the process observed. The facilitator and the champion, after directly
observing the targeted process, should be able to determine what the base
target for improvement will be. Also included on the charter document are
the names of the champion, the facilitator, and the event duration. The charter is now complete and it is time to move to the next step in the planning
process: selecting the team members.

Kaizen blitz events are team-based events that take advantage of the
power of teamwork. Teams bring diversity in skills, thought, and knowledge.
Inviting individuals to join the team should be based on some underlying
reasoning. Some examples include:

Kaizen Blitz Team Charter

This document is used to define the scope of the team's efforts.

Kaizen definition: Japanese for "change for the better." English translation is "continuous improvement."

Goal: The elimination of waste (defined as "activities that add cost, but do not add value"). A closer definition of the Japanese usage of kaizen is "to take it apart and put it back together in a better way." What is taken apart is usually a process, system, product, or service.

Process title Identify the process that will be observed and improved in this section.		
Process Boundaries		
Starting point – Determine the process boundaries— starting and ending, and note them here.		**Ending Point -**
Key Objective – In this section, note the objective and the metric target.		
Champion – Identify the leadership individual who requested the kaizen event.		
Facilitator – Identify the facilitator who will play a content neutral role and guide the team.		
Event Duration – List the dates and times for the planned event.		

Figure 5.4   Kaizen blitz team charter. This document is used to define the scope of the team's efforts.

- A process expert or two who works daily in the area to be improved.
- People who have a passion for continuous improvement.
- An individual who is known to be negative about the topic of continuous improvement and will give you the opportunity to change his/her thinking during the event. This is my favorite person on the team. As a facilitator, I give him/her special focus to ensure he/she will be converted into a champion of continuous improvement. After the kaizen event, this former nonbeliever will help to change the company's culture.
- Individuals working in one of the upstream or downstream operations that can bring those insights to the team.
- People possessing technical skills you may need during the event. Examples include someone from your systems department if the team is focused on a paper flow process or a maintenance technician if you think you might have to move or relocate equipment or power lines during a shop floor event.

Part of a facilitator's job is to ensure he has staffed the team appropriately to help ensure a successful event. For that reason, in addition to considering the skill set an individual may bring to the team, the number of individuals invited to join the team is just as important. Required is enough staff so that they can accomplish the team's goal, but not too many. Having the team overstaffed could result in individuals who do not fully contribute to the team effort. Since kaizen events are "hands-on getting dirty together" events, the last thing wanted by the facilitator are people standing around with no opportunity to contribute. Teams of five to eight members is the norm, but the staffing level can increase based on the complexity of the process observed. Later, I will describe some specific roles assigned to team members during an event. A facilitator requires enough people to fill these roles, but not too many so that individuals do not have a defined role. Experience is probably the best teacher when it comes to team selection and staffing. Following this team selection step is the requirement to complete some detailed planning and scheduling activity.

The first step is to send an invitation to the individuals selected to join the team. A written invitation, addressed to each individual, is appropriate because it conveys the importance of the event along with the importance of this individual to the team. The letter should include a brief description of the event, the timeframe for the event, and a requirement for them to seek approval from their supervisor and work team before accepting the invitation. They are usually given two to three days to respond and are offered the opportunity to stop by my office before responding with a yes or no, so that I can answer any questions or concerns they may have about the event. Step two is to make arrangements with the process owners to ensure equipment or process availability during the event. For example, if you are going to be observing an assembly operation, you need operational approval to interrupt their operation. They may elect to work some overtime the week before the event to build in some extra capacity for your team activity. Step three in detailed planning is to reserve a meeting room for the duration of the event. Kaizen events are formal processes and the team should have a defined place to meet and work. It lends credence to the importance of the event. Next step is to notify any support staff whose support you may need during the event. For instance, informing the maintenance supervisor that, during the event, the team might want to move equipment around in the work cell to improve cycle times, or reduce ergonomic injury risks. This should be completed well in advance of the event. This allows the maintenance supervisor to do some resource planning so that he can support your request as soon as the team requests the moves. As a facilitator, one of your roles is to ensure all activities happen rapidly during the event. Because you are using the allowed time to complete the kaizen event as a means of pushing the team to action, you cannot allow for any delays. A maintenance team response that they will get to it tomorrow is unacceptable. One of the values of holding a lean safety kaizen event is that the focus is safety. Safety work orders are generally a priority for the maintenance personnel, so it almost guarantees a quick response.

Everyone appreciates a free meal, thus the next step in detailed planning is just that. Gathering some menus and securing resources to order and deliver lunches daily is the task. A free lunch each day of a kaizen blitz event is an important part of the team building these events require. Whether it is a working lunch or not, it is an opportunity for the individuals to get to know each other better. I often start the conversation by asking about their interests outside of work. Once you get the conversation rolling, others will jump in and, before you know it, the lunch period is over. Lunches need not be elaborate or expensive. Pizza, salads, soup, and sandwiches or other choices will do. It is a low-cost way of building the team. To find out about a person's passions, what really interests them, and to get to truly know them, it is important to get them to open up about their interests outside of the workplace. They will gladly share them with you over lunch once you initiate the conversation. I can tell you from my personal experience, these lunchtime conversations help me to really connect, and build a lasting relationship, with all the individuals on the team. Take the time to get to know everyone on the team as the individual they are. Life is so much more than work and they have stories and life lessons to share.

An important part of a kaizen event happens after the event has been completed. This includes a presentation by the team to others, both about the team process they followed and the results of their event. We will cover that presentation process later, but at this point it is important to decide on the attendees and reserve a meeting room for the presentation. Included in the invitation should be the business leadership team, so this is a chance to check and then reserve the meeting time on their schedules. The last step until just before the event takes place is to compile any training material that will be delivered the first morning of the kaizen blitz event. Selection of the material will depend on both the focus of the kaizen event and the lean experience of the participants. If the team has quite a few lean novices, included in the training material should be an overview of lean terminology and the materials on the seven wastes. Whereas, when the focus is safety and the reduction of ergonomic risks, some basic training in ergonomic assessment techniques is required. People cannot be expected to identify waste or improvement opportunities in a process if they do not know what they are looking for; therefore, up-front training is very important. Planning the training well before the event starts allows for the selection or development of the appropriate material and time to practice the presentation of the material. The initial detailed scheduling and planning for the kaizen blitz event is now complete. The next suggested activity should take place a few days before the event begins, which is to send out a reminder to everyone who was invited to participate. Include the start date, time, and meeting location in your reminder.

Before the team arrives on the first morning of the event, the room has to be set up and the supplies required by the facilitator and the team made available. In addition to the obvious tables and chairs, flip charts, markers, Post-it® notes, clip boards, kaizen blitz forms, and training packets are some of what might be required. When the team arrives, it is a good practice to consider some method to put the attendees at ease. If they have never before participated in a kaizen event, they will be a bit nervous, so by simply asking them to introduce themselves and mention their favorite pastime outside of work you can put them at ease. Figure 5.5 is the next process map we will review. It covers the activities of a multiday kaizen event.

Kaizen Blitz Step 2 - Facilitating the Event

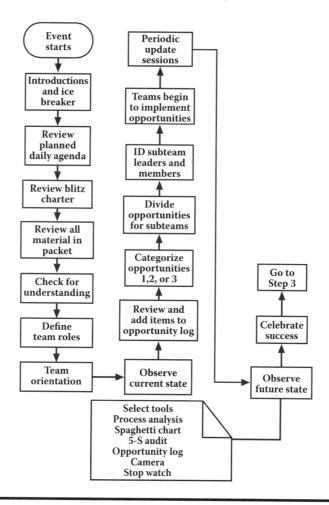

Figure 5.5   Kaizen blitz map 2.

Facilitating the Kaizen Blitz

After the short icebreaker to calm the attendees, the facilitator can provide an overview of the agenda for the multiday event. The agenda provides the attendees with a road map of their planned activities, which will also help to put them at ease. For a three-day kaizen event, the agenda would look something like this.

Day 1
>7:00 start time
>Introductions, training, review of event objective

Team member roles assigned and required materials distributed
Go to the work area, observe the current state process and identify
    opportunities
Lunch at 11:30–12:15
Review opportunities and gather additional improvement ideas
Divide team into subteams and split the opportunity list between them
Subteams begin to implement agreed upon changes

Day 2
7:00 start time—subteam leader report out
Subteams continue to implement changes
Lunch at 11:30–12:15
Subteams continue to implement changes

Day 3
7:00 start—subteam leader report out
Continue to implement changes and prepare to observe process second
    time
Lunch at 11:30–12:45
Observe future state process and record any new opportunities
Celebrate success and begin to plan team presentation

This schedule overview is a rough draft. To provide a clearer understanding of the facilitation steps and the reasoning behind them, I will review the
process map and provide more detail as if this were a lean safety kaizen
blitz event.

First is a presentation and review of the team charter, which was drawn
up by the champion and event facilitator. This will be the first time the team
has viewed the document, so give them time to read and then ask questions. Remember, a transfer of ownership is critical if they are to feel the
objectives stated on the charter are to become their objectives. Don't be surprised if they do not share the same level of excitement as those who created the document when they first read it. This should be expected. I have
been involved in many management meetings where changes are proposed,
debated, and after multiple meetings on the topic agreed to. Then when
they are presented to the employee base, there is a negative reaction. This
is natural because those in the audience have many of the same questions
and concerns as the management team and should be allowed the same
opportunity to question and debate the changes as those who developed the

proposal. A leadership lesson is to only present the changes in a very brief initial meeting. State very clearly that because this is the first time they are learning about the changes any questions they have will be addressed in a second scheduled meeting a few days later. Note that you want to give them time to absorb and think about the changes. When the second meeting is held, much of the emotion evident in the first meeting will have dissipated because they were given time to think and talk about the changes presented amongst themselves and with their supervisors. The kaizen team's charter should not cause that type of emotion especially if they are being asked to improve safety. Remember that the charter will provide information of the process they will observe and improve along with the process boundaries and the improvement objective. Since this is a lean safety event, their objective could be to identify and implement 15 safety improvements that will reduce the injury risk in the process observed. After sharing and discussing the charter, the next step is to review the balance of the material in the packet that was compiled for each attendee. This is the time to review and present the training materials that were included.

Because safety is the focus, and the team will be asked to observe and improve the work methods in order to attain their goal of 15 safety improvements, ergonomic assessment training is appropriate. Ergonomics seems to most like a mysterious scientific approach used to analyze work. Here is an overview sourced from Wikipedia that supports that thinking.

> Ergonomics is the scientific discipline concerned with designing according to human needs, and the profession that applies theory, principles, data and methods to design in order to optimize human well being and overall system performance. The field is also called human engineering, and human factors. To assess the fit between a person and their work, ergonomists consider the job being done and the demands on the worker; the equipment used (its size, shape, and how appropriate it is for the task), and the information used (how it is presented, accessed, and changed). Ergonomics draws on many disciplines in its study of humans and their environments, including anthropology, biomechanics, mechanical engineering, industrial engineering, industrial design, kinesiology, physiology and psychology.[1]

The topic of ergonomics and MSD (musculoskeletal disorders), or repetitive strain injuries, over the past eight years has been high on the list of topics for every safety program in every plant. Beginning on January 16, 2001, OSHA's Ergonomics Program Standard for "general industry" was slated to become law. There was a lot of debate about the pros and cons of this legislation and in the end OSHA backed off on this change and has since taken an industry by industry approach to defining ergonomic standards intended to eliminate or reduce MSDs. One thing this action did accomplish was to raise the awareness of ergonomics. Employers, employees, industry specialists, doctors, and lawyers all became involved. Ergonomic-related soft tissue injures, such as carpal tunnel syndrome, became part of the common vernacular, when prior to 2001 most had never heard of it. At the manufacturing plant where I worked, in preparation for the proposed OSHA enforcement date, an ergonomics team was formed and an industry specialist was brought in to provide ergonomics training. I still remember the three-inch-thick binders the trainees carried and the dazed looks on their faces after the multiday training concluded. They were given way too much technical information to go out and make a real difference. So, how does the ergonomic assessment training material that should be provided to the kaizen blitz team compare?

The training for the kaizen team consists of four words or statements and a brief description for each. They are as follows:

1. *Out of neutral*: Whenever the team observes an individual working and one of their body parts gets out of the neutral position, it is an opportunity for improvement. Out of neutral examples I usually provide include someone's arm above their shoulder. That means the shoulder is out of a neutral position. When someone has to reach for something and lean forward, his/her back goes out of neutral. If you have to twist to get something, your torso is out of neutral. These are simple examples that make the concept easy to understand and grasp.

2. *Straining*: Whenever someone has to strain himself to exert physical force to an object, it is an opportunity for improvement. An example from home is trying to open a jar of pickles. There are better examples in most facilities.

3. *Heavy objects*: If an individual has to move or lift a heavy object, or get help from someone to do the same, it is an opportunity for a safety improvement.

4. *Repetitive actions*: Anytime someone is asked to repeat short duration tasks repetitively, it is an opportunity for improvement.

That is the extent of the ergonomic training the kaizen team will receive prior to observing the work process to be improved. In the beginning, when the team begins to observe the work process, the facilitator can point out one or two of these conditions to reinforce what the team will be looking for. No science or measurements, just the basic kaizen approach of direct observation along with a few general guidelines to follow. Anyone can identify unsafe conditions, or at least targets for safety improvement, by following these simple, yet effective, guidelines. These methods are not intended to minimize or make light of the skills and talents of trained and certified ergonomists. When you want to get into the detailed analysis and measurements sometimes required for workstation design, a licensed ergonomist should be consulted.

Other training materials that would be included in the packet and reviewed with the team would be related to the lean terminology and lean tools that may be used during the event. As the facilitator refers to words, such as work cell, flow, load leveling, etc., the team should have a general understanding of the "lean" meaning of these terms and how they apply to their activities. That is another benefit to a company's lean program. The team starts to learn and understand the lean language as they focus on safety improvement. As the facilitator reviews the training materials, it is always a good idea to check for understanding by asking the team questions like, "Could you repeat, in your own words, what load leveling means?" When the training has concluded it is time to assign the team members their roles.

As I have repeatedly stated, a kaizen event is a great opportunity to engage and grow people. When the facilitator and champion selected the team members, they did so with forethought and intent. Now is the time to take advantage of that earlier selection process and assign team members roles that will allow them to contribute to the success of the team and at the same time give them the opportunity to grow. The roles will vary somewhat depending on the focus of the kaizen blitz team. Because this team's focus is safety, the number of roles having to be filled is less than a lean-focused kaizen team that is charged with reducing the cycle time of some work process. The reason for this is to make certain the team's focus is on safety. Therefore, the traditional tasks of manning a stopwatch, to record the cycle time of each process step, and maintaining a process step log are no longer needed. The roles remaining are for individuals to complete the following assignments:

Opportunity Log—Someone will capture any and all potential improvement ideas the team identifies on an opportunity log. Figure 5.6 is an example of such a document. Idea generation during a kaizen event is treated

Kaizen Blitz Opportunity Log

Date: _____   Process observed: _____

No.	Opportunity	Subteam Name	*Priority 1,2,3	Comments
1				
2				
3				
4				
5				
6				
7				
8				
9				
10				
11				
12				
13				
14				
15				
16				
17				
18				
19				

Use this form to gather all ideas while observing the current state process. Use brainstorming guidelines, e.g. there are no bad ideas!

*Note: Priority definitions – 1 means the team will complete the task during the event, 2 signifies that the team or someone else will have to follow up after the event and 3 denotes a capital investment is required and the team will not pursue the opportunity.

Figure 5.6   Kaizen blitz opportunity log.

like brainstorming. The facilitator will urge everyone to shout out all ideas so they can be recorded. Select an individual who can initially clearly write down the information and then later enter this information into a computer in order to turn this list of ideas into the team's task list.

Photographer—Ask someone to take digital photos of the current state process, the team activities as they work to improve the process, and, finally, the future state process after the changes have been made. These photographs can be used as part of a PowerPoint® presentation prepared for the report out meeting held after the event is completed. If someone on the team has an interest in photography, it is a chance for him/her to display his/her talent. Or, if there is an individual who has no digital camera experience, it may be your chance to mentor someone by encouraging them to take the task.

Process expert—This is the individual whose regular job is to perform the work tasks that will be analyzed and then improved. They are the expert and bring a lot of knowledge to the team. Despite the fact that the team will focus on the work process and not the person, the process expert can sometimes be resistant to the suggestions for change for they may have performed the same tasks for a number of years and often feel their way is the best way. A good facilitator will find a way to ensure their ego is preserved as the team explores and implements new ways to perform old tasks. Another benefit of focusing on safety is that any perceived or real resistance should quickly fade as the process expert realizes the team is making the tasks to be performed safer and easier. An example of a situation like this occurred recently when I was facilitating an accident investigation. An individual, who took great pride in training new sales staff in the application of a product, was diagnosed with tennis or golf elbow despite the fact that he didn't play either game. What he did have to do was repetitively swing a hammer along with other repetitive actions as part of this trainer position. We took a minikaizen approach during our investigation. Four of us went back to the individual's work area and directly observed him perform his tasks. As he was working, he explained his work steps and why he performed them in the manner he did with a sense of authority and ownership. He was the expert and he wanted us all to be aware of that fact. As he proceeded, the four of us started to build a list of opportunities to reduce the ergonomic-related risks inherent in his work. As we did, we also started to earn his trust. It took about one month for all the changes in his work area to be completed. Twice during that period he stopped me in the plant and asked if I had seen the latest changes. When I replied that I had not, he invited me to join him so he could show off the changes. He had changed from someone who was unsure about, and maybe even a little resistant to, change into a champion for the changes that were made. The value of the changes relative to his safety became evident to him and he was won over. A facilitator has to invest time and energy to win over the process expert on any kaizen team. They are one of the keys to the team's success.

Idea generators—Everyone on the team fills this role. All are expected to identify opportunities for improvement as they observe the current process. Having this generic role accomplished two things. It makes everyone feel that they have a role and it allows the facilitator to challenge them to fill the roll. More traditional kaizen events, that focus on process cycle time, have many more roles for individuals to fill. Some of them are process step document recorder, a mapmaker who will draw spaghetti diagrams, and a time keeper who will track process step cycle times using a stopwatch. During safety kaizen events these roles should not be filled to ensure the focus is on safety, not cycle times.

Now that everyone understands the role they will play, it is time to take the team to the work area where the kaizen team will observe the targeted work process. Notice that this step is called *team orientation* on the process map. This is the facilitator's opportunity to ask the process expert to both brief the team on any safety hazards they should be aware of, and to introduce the team to their co-workers in the area. The co-workers should be gathered together and briefed on what the team's object for the event will be. Keeping all the workers in the area informed of the team's actions will help to guarantee their acceptance of the changes to come.

The stage has been set and it is now time for the process expert to perform the work to be observed and improved by the team. The facilitator should make sure everyone understands his/her roles and then signal the start of the current state review. An experienced facilitator will call out a few potential improvements to salt the pot and get the team to more closely observe the work being performed. After the work task has been finished and all ideas gathered, guide the team back to their meeting room. Ask for someone's help to complete the next step on the process map, which is to review each idea that was generated. Let the volunteer read them one at a time while the team is challenged to dig deeper to improve existing ideas or add new ones to the list. When this task has been completed, it is time to have the team divide the opportunities into three categories. All opportunities will be given a 1, 2, or 3 ranking. The meaning of these priority designations are as follows:

1. These are opportunities the team believes they can implement during the three-day event. A skilled facilitator will push them a bit to get a 1 rating on as many opportunities as possible.
2. Opportunities that may not be completed during the event and will be assigned to the local area supervisor to follow up on following the event.
3. Any opportunity that requires capital spending is given a 3 ranking. Management will have to determine, based on an ROI (return on investment) analysis, if they want to pursue the opportunity after the event ends.

While the opportunities are being reviewed and the priority codes assigned, the facilitator should be mentally categorizing the opportunities into two groupings based on similarity. The two categories might be all tasks related to the workspace layout and the second being material handling improvement tasks. Meanwhile it is time to ask for two volunteers to be subteam leaders and then split the tasks between them. Discourage individuals who already have leadership roles in the business from volunteering. Instead seek individuals to grow and develop. One of the joys of facilitating kaizen blitz events is watching people grow. Do not pass up this chance to encourage individuals who might not normally volunteer. Occasionally, they may have to be directly asked if they would be willing to give it a try. Let them know they will be successful because they will receive mentoring and support to ensure their success. Once two volunteers are in place, even though—if they were pushed *hard* enough—they may feel they were "voluntold," it is time to divvy up the rest of the team in order to form two subteams. This is an important step because there are many tasks to be completed and only 2 ½ days left to accomplish them.

Now release the two teams, with their opportunity log tasks in hand, to begin their research, testing, and implementation of change. It is now time for the facilitator to step back and become less involved. If the subteam leaders are to grow, they must be allowed to lead. Simply check with them occasionally to ask if all is well or if they need your assistance. Gather the two subteams about twice a day, start of shift and before departing for the day works well, and allow the subteam leaders to give an update on their team's progress. By early afternoon of the third day of the event is the time to stop making changes and once again observe the work process. This time the team will observe the "future state" process to determine the impact of their improvements. Based on my experience, a safety kaizen event yields an average of 15 to 20 safety improvements and the team has a vested interest in this second pass process review being successful. Team building can be a long laborious process. Anyone who has attempted to form a group of varied and unique individuals into a cohesive team understands the difficulty of the task. Kaizen events on the other hand have the power to build a cohesive and focused team in just three days. A few years ago when I facilitated a kaizen blitz sponsored by the Association for Manufacturing Excellence (AME), I learned firsthand how strong this sense of teamwork can be after only two days.

The team had been charged with reducing the changeover time on a metal stamping punch press by 50 percent. They worked diligently for two days to implement all of the changes they had indentified during the first pass, or current state, changeover. Midmorning of the third day, we were going to watch the process expert, the press operator, complete the same changeover and record the new cycle times to see if the team had made the 50 percent improvement goal. One small problem was immediately evident: the process expert was not there when we gathered in the meeting room. About 15 minutes later, he arrived looking like he had a very rough night. He explained that he had decided late in the evening to go to a hospital emergency room to have a large cyst removed from his neck. He did not get home until about three o'clock in the morning and had considered staying home, but he did not want to let the team down. Though we encouraged him to go home, he insisted on staying to perform the changeover. In the middle of the event, the dressing on his neck had to be replaced, but he returned, finished the changeover, and the team's goal was attained. His dedication to the team taught me about the power of kaizen blitz team-based events. The lesson learned is that, to successfully build teams in a work setting, they must have a strong common focus or goal. Without that element present, they are teams in name only. Kaizen blitz events provide a common goal that helps to build and then bind the team together. The press operator for this kaizen event taught me that lean leadership lesson.

Now, back to our kaizen team. After they observe the second pass or future state work process to determine the impact of their improvements, they will be very proud of the results. What is remarkable is that they will not only recognize the safety improvements they have made, but they will also comment on and be proud of the cycle time gains that have resulted from their safety event. In two recent such safety kaizen events I facilitated, the team reduced the labor content required for the work process by 50 percent. How is that possible? By making safety improvements mainly related to improving or eliminating material handling steps in the process, the team accomplishes the dual goal of safety and cycle time gains. It works every time. The facilitator should guide the team back to the meeting room and use this time to recognize the team and each individual for his/her special contribution to the team's success. During the event, I make mental notes of what I want to say to each individual. They accomplish so many wonderful things during the event that the difficult part for me is selecting the most appropriate action to recognize. I believe the greatest motivator is accomplishment. This is the perfect time to recognize their accomplishments. I also firmly believe that the biggest benefit of any kaizen event (it matters not if the focus is safety or cycle time) is the minds you change, not the process you have improved. These five to six people are changed forever. They are slightly infected with "lean fever." So, let's move to the next process map (Figure 5.7) that describes the planning for and the presentation of the event results by the team.

Kaizen Team Presentation

Many of the team members will have never had the experience of getting up in front of a group of managers to do a presentation. This is a scary proposition for them, but an opportunity for the facilitator to continue to engage and grow the team. Helping them to plan this presentation, rehearsing the meeting when necessary, along with constantly reinforcing that they will be fine and will do a great job will help to prepare them. The facilitator should schedule the meeting time and place and send invitations to the appropriate people. Attendees can include the management team, co-workers of the presenters from the area of the process change, and any upstream or downstream process owners who may benefit from the changes made. This is a cultural impact meeting—a celebration of the team's success. It is a marketing tool to help drive culture change. Do not underestimate the power of having peers present their continuous improvement story to their co-workers. They will have a much more attentive audience than any manager can ever hope to have. In smaller businesses, it is appropriate to have the entire workforce in attendance.

To prep the team, it is wise to mirror the same sequence of process steps for the presentation that were used in the three-day event. This gives them confidence because they have just followed this process for three days. Each of these steps, or talking points, will have a team member assigned to it:

Kaizen Blitz Step 3 - Presentation

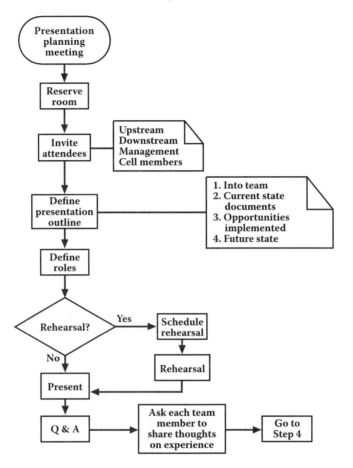

Figure 5.7   Kaizen blitz map 3.

■ Introduce the team
■ Review the team charter including goal
■ Describe the current state condition
■ Overview the improvement opportunities identified
■ Review some specific changes that were made
■ Describe the future state condition and the benefits recognized
■ Answer questions from the audience

As this presentation planning takes place, the facilitator should take notes and then prepare the meeting presentation. I most often use PowerPoint and include many of the event pictures in the presentation. The team will be nervous, but they all survive and grow from the experience. The facilitator should only play a role at the very end of the presentation. This is when I ask a simple question to which all team members must respond. They are always informed ahead of time that I will ask it and are told to think about their response. The question is: "What did you think about being given the gift of time to participate on the kaizen team?" Their heartfelt, honest responses are usually the highlight of the meeting for me. As I listen to them talk about their personal growth, I watch the faces of their senior managers. This is the evidence that the culture has been positively impacted and that I have connected with them. As a final step, when possible, the team should lead all of the meeting attendees out to the work area to demonstrate the changes they have made. This "frosting on the cake" step is another highlight for the team and the meeting attendees. I usually focus on the senior leaders faces to observe the same pride in their workforce that I feel.

The event is over and everyone goes back to his or her regular jobs, but there are often open tasks that still need attention. Figure 5.8 displays the steps to ensure follow-up occurs.

Kaizen Event Follow-Up Meeting

As noted on the chart, the facilitator schedules a follow-up meeting that includes the team and the supervisor of the area where the kaizen event took place. All tasks are reviewed and target dates updated. Ownership of the task list changes from the facilitator to the supervisor at the end of this meeting.

As I stated earlier, the kaizen blitz is the most powerful people engagement, cultural impact tool in a lean thinker's tool kit. I hope this kaizen overview that resulted in safety improvements, cycle time gains, and people growth helped you better understand the power of this tool to drive change. Now, let's look at the application of all of the theory, philosophy, and benefits of the kaizen blitz in an actual kaizen event held in a small manufacturing plant.

The company I worked for is part of a self-funded worker's compensation trust organization. The trust is composed of about 100 small- to medium-sized companies and is managed by a management firm. Each trust member site would receive multiple visits each year by a loss control representative, from the management firm, who would conduct safety audits and share his/her safety expertise in many ways. The trust was and is a great organization bound together by a passion for safety improvement. A few months after a job change, in which my role shifted from operational responsibility to a full-time lean office leadership role, I was talking with our loss control

Kaizen Blitz Step 4 - Follow-up

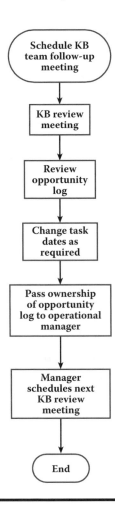

Figure 5.8   Kaizen blitz map 4.

representative who was visiting to follow up on some items discovered during an earlier audit. One of our corporate values, integrity, encouraged us to "be a positive presence and contributor to the community." I had been thinking about how I could, given my new lean role, use my lean talents and knowledge to fulfill that responsibility. What I proposed to him was that I would facilitate, at no cost, a kaizen blitz event at another member site. His role would be to locate a site that would host a team composed of some of their employees plus outside attendees from other member sites. My broad goal was to reduce the trust's overall workers compensation costs by taking lean thinking principles and applying them to safety within the trust. The trust, in its search for new member companies, wanted to pursue and enroll new companies who were on the lean journey. I thought why not take the companies already in the trust and use safety-focused kaizen events to both improve safety and start or restart their lean journeys. I envisioned four to five member companies represented on the kaizen team so that I could touch more than just the host site. The category of injuries that were dominating the claims paid by the trust was ergonomic-related, soft tissue injuries. To me, the kaizen approach seemed like an obvious solution. After explaining this to the loss control representative, he looked at me and he said, "What did you say?" Kaizen blitz and other lean terms were words from a foreign language to him, so I had to slow down, curb my enthusiasm, and explain in detail what a kaizen blitz was and how this type of an activity might impact safety long term in the trust. In the end, he agreed to try and locate a site that would consent to sponsor the first kaizen event. It took a few months, but he eventually found a volunteer site.

A planning meeting was scheduled at the volunteer site. The agenda included meeting the management team and reviewing its operational processes in order to select the targeted process that would be observed and improved during the kaizen event. The event was advertised, via a flyer and emails, to all of the trust membership offering positions on the kaizen team at no cost to their company. Within a month, the team roster was filled with representatives from five trust sites. The event was held and was more successful than I had envisioned. It was held at a manufacturing plant with around 150 employees. It was the first ever kaizen blitz event in the facility. The team's efforts yielded about 15 implemented safety improvements along with reduced cycle times and improved throughput. All of these improvements resulted from observing a finished goods packaging operation. The young lady who worked at this packaging station was not at all thrilled on the first day when she had 10 people watching her work and challenged the work process each step of the way. She came back and endured us on Day 2 and by Day 3, she was the main team cheerleader. During the presentation to management, she publically stated that she had thought in the beginning we were wasting our time, but that she had been wrong. She couldn't wait to take everyone out to her new workstation to demonstrate her now safer and more productive work process. When the team had completed their presentation of results, she led the management team to the shop floor for a demonstration of the new process. After observing this demonstration and seeing the passion and acceptance of change in his employees, the business owner leaned over to his chief engineer and said (*paraphrasing*), "We need to do this all over the plant." One of the guest kaizen team members was also very impressed because he went back to his facility and shortly thereafter called the trust's management firm and offered to host the second of these safety-focused kaizen events in his facility. The second event led to a third and the third led to a fourth. It is this fourth event that I want to cover in some detail.

The site that hosted the fourth safety kaizen blitz was a small company with around 50 employees. They were members of the trust for many years and displayed in words and actions a strong focus on safety. As evidence of their safety first culture, they had recently surpassed three years without a lost time injury. Considering the type of work that was performed, this was an accomplishment of which to be proud. It was a second-generation business that supplies metal parts to numerous differing industries. They were a job shop. Job shops often have difficulty relating to lean principles because its work is often varied, therefore making it difficult to see a regular "flow" through the shop. It is as if they are saying, "If I am doing something different every day, and I never know what that will be day to day, how can I improve it?" This job shop was no exception. Other than some rudimentary application of 5S, there was little visible evidence of lean in the plant. However, one thing was different here. The owner believed in and understood the value of lean to his business. He had hired lean consultants to provide lean training for some of his staff and yet expressed dissatisfaction with the lack of progress made to date. I was confident that the lean safety approach of focusing on safety improvements would result in a lean foundation upon which the owner could build.

In preparation for the three-day kaizen event, I made two visits to the site along with the trust loss control representative. We were given a tour of the plant and had a chance to meet and talk with some of the staff. As with many first- and second-generation businesses, the culture was family oriented with the owners firmly in charge. As we walked around during the first visit, we posed many questions to help us understand both the product flow and the operational steps that were being performed. We were looking for processes that required numerous material handling steps that would provide the team with ample opportunities to positively impact safety. Since they produced parts in batches, it didn't take long to decide upon the operational process the kaizen team would tackle. It was a three-step process that formed, assembled, and welded a sheet metal fabrication. The event dates were selected and then advertised by the trust. Two different member sites offered to help staff the team. Having some external eyes on a kaizen team is very important because they will challenge the status quo. The host site selected five individuals from their staff to complete the team. The team of eight, plus the loss control representative and me, were ready to begin our team journey together.

We began in a small meeting room with almost everyone looking a bit nervous. To help put them at ease, we began by having everyone introduce themselves and their work responsibilities. The external team members represented two companies. One of them was a manufacturing manager and the other two, from the same plant, were frontline supervisors. The internal team members were mostly from the plant floor. Included were a supervisor, a lead man, and two operators. The final member of the team was the sales engineer for the company. His everyday role was to make sales calls to the company's varied customers. I had been informed ahead of time that he had been selected because he was a little negative when it came to the topic of "lean." There is nothing I like better than having a naysayer on the team. I make it my personal goal to change their thinking in three days. Once the introductions were complete, we moved onto the PowerPoint slides that had been prepared ahead of time by the loss control representative. The team, for the first time, was made aware of their objective. They were to reduce MSD injury risks associated with the operational process we had earlier selected. They were informed that, although this was a lean event, our focus would be safety. Since it was a lean event, we continued the training by introducing them to the lean terminology they would be hearing over the next three days. Safety is a language they all knew very well, but lean was new to most. That is another benefit of these lean safety events. You gradually introduce lean while focusing on safety. This approach makes it less threatening and more acceptable to those who do not understand the true meaning of lean and have only heard that is a cost-savings method employed by management. When the training was completed, the internal staff overviewed plant safety for the visiting team members, issued us the safety apparel required, and we proceeded to the shop floor.

Our team charter had defined our boundaries, so we started with a flat sheet metal part that has already been stamped from coil stock at a preceding operation. The team's first stop was a machine that bent up all four edges of the blank sheet metal to form a pan. What the team observed was that the parts were brought to this machine in a container that required the operator to bend over to retrieve the vertically stacked parts. The operator also had to pry apart the sheet metal pieces with his fingertips before he lifted them from the container. The oil that had accumulated on the parts from the previous stamping operation caused the parts to stick together. The team immediately recognized these two safety concerns, bending and prying parts with your fingertips, as opportunities for improvement. Next, they observed that the operator, as he was rotating the part in the machine that formed the four edges, had to wear wrist restraints. The type of equipment used to bend the sheet metal is very dangerous and there are OSHA rules that govern the types of safety devices that can be used to ensure operator protection. These wrist restraints pulled the operator's hands back just far enough when the machine cycled, triggered by the operator pressing a foot control pedal, to make it impossible for the operator's hands to be in the tooling danger zone. Wrist restraints work very well to protect the operator, but the team noticed they also interfered with the operator's ability to move freely while performing the required job tasks. This was made very clear when he had to try and put the completed part into a corrugated box and the wrist restraints limited his range of movement. Another opportunity was added to the list. The team also noted that the corrugated box was not quite large enough to hold the parts, which caused the operator, while bent over, to manipulate the part to make it fit. Another opportunity. After they had observed about five more parts being processed, they watched the operator look for and retrieve a pallet truck to move the corrugated box full of parts, which was on a wooden pallet, to the next operation. They identified this as an opportunity because they observed the effort to move this load across a floor that was not always smooth.

The next work center included a spot welder that was used to weld four smaller sheet metal parts to the larger part. Two operators staffed this work center; one assembled the small parts, using a fixture, to the large part, while the second operator manned the spot welder. The team first noticed that the small parts were thrown haphazardly into storage containers and those containers were on the floor. This required the operator to bend over to retrieve them and then, using hand and wrist motions, orient each part into the fixture. Two more opportunities were written on the log. After the parts were spot welded together, the first operator moved the part to a holding fixture, which would release the welding fixture from the welded assembly. He then grabbed a rubber mallet and proceeded to conduct a quality test on the part. Using the rubber mallet, he offered up multiple blows against each of the four small parts that had been spot welded to the base to check the quality of the spot welds. The part was stationary while it was being hammered, so the operator had to rotate his wrist to hammer in four, sometimes, awkward directions. The team picked up on this and added another opportunity to improve safety to their log. After the hammer test, the parts were again put into a corrugated container by bending over and placing the parts into the box. They had seen this action before and knew they needed to address it. Then, after the pallet truck was retrieved, the operator had to physically move them to the next operation. The next operation, welding, was at the opposite end of the plant, thus the physical effort involved resulted in yet another opportunity.

As they gathered around the welding station, they observed the welder, who was a member of the team, bend and reach into the container to retrieve a part. This was the third time they observed this action. He then located the part into a fixture and welded a hub to the center of the base. You can guess what he did next right? He bent over to put the part back into the box. Directly observing a process with an eye for safety improvement had turned up many opportunities. It is the same all over the world. People work in the way they were trained or "as it has always been done." Giving people the gift of time to directly observe and improve something was going to pay big dividends for this company. The team now had a chance to review all of the opportunities they had identified and decide on their action plans.

They were directed back to the meeting room where each item on the log was assigned a 1, 2, or 3 code as described earlier in this chapter. As the list was reviewed, the team came to some decisions, such as agreeing to ask management if they could move the three operations together to eliminate both the physical material movement around the plant along with all of the steps related to putting the parts into and then taking them out of containers. In the end, the owner agreed and the maintenance staff was immediately employed to begin planning the move of the spot welder and the welder to the first workstation where the parts were initially formed. In order to complete all of the tasks on the log, the team now had to be split into two subteams and the tasks split between them. I always encourage internal participants to take these leadership roles and two of them did. The shop floor supervisor and, to my surprise, the sales engineer stepped forward to lead the subteams. They both did a great job of identifying, testing, and implementing changes with their subteams over the next two days. The results, which soon will be described, speak for themselves.

By late morning of the third day of the event, the team was ready to observe the future state process as they had set it up. They now had a work cell rather than three separate work centers between which parts had to be transported. They aligned the equipment so that the part was not put into a container until it was completed and ready for shipment. The process began with the blanks being delivered on a pallet in a horizontal position. They had attached a sheet-fanning device at the edge of the stack. This device uses magnets to fan or lift one edge of the top sheet away from the stack. The operator now, while remained vertical, picked up the first sheet without the finger prying previously required. He now was no longer restrained by the wrist restraints either. This operator, in the newly defined work center, was required to weld the hub in the last operation and that would not have been possible if he was strapped to the first machine. The team identified that the two-hand control console, which was already part of the machine, could be used rather than the wrist restraints. They also observed that the operator, in the current state, had to push against the sheet metal while it was being gripped by the forming tooling. This was to ensure it did not change locations resulting in a quality defect. They had to find a solution that would allow the operator to let go of the part. They implemented a great idea, which was to have magnets inserted into the machine tooling that would hold the part in place and free the operator's hands to press the two-hand control that would trigger the machine to activate. Having both hands on this control panel ensures operator safety just as the wrist restraints had. The team also noted that the spot welding operation had the longest cycle time in the entire production process. It was near or equal in time to all of the other operations. As a result, the same operator, staying in the same location, affixed the fixture and the four small parts that would be spot welded. The small parts were now in plastic bins and all oriented in the correct direction. After completing this step, the part was passed to the second and only other cell operator who was positioned at the spot welder. As he was spot welding the part, the first operator put on his welding helmet and welded the hub to the last part that was spot welded. When he finished, he moved the part to a container that had been tilted forward and lifted to reduce the bending and reaching previously required. He then took off his welding helmet and went back to the first operation and began to bend the edges on the next blank. They had constructed a lean one-piece, flow work cell by making many safety improvements.

The staffing level in the cell was two operators. You may remember in the current state process, two people had been assigned to just spot welding. Although we never used a stopwatch, in the middle of the event when the equipment was being moved together, the team asked, "How many people are going to work in this new work cell?" As all facilitators do, since they have to have the right questions, not the right answers, I responded, "How many people do you think you will need?" The answer from one or two of them was maybe only two. I responded that the only way you will find out is to set up the cell and then run test parts through the process. Well, they did and the results were as described above. They had hit a lean home run as a result of their safety improvements. The team's next step was to plan their presentation.

While the loss control rep worked feverishly to download photos from the event and pull together the PowerPoint presentation, I prepped the team. Each member was assigned a role during the presentation, which was to start in less than two hours. Rather than present to just the management team, all the employees were gathered together in the cafeteria to hear from their peers about the first-ever kaizen blitz event in their plant. The team stood in front of everyone and told their story following the standard script described earlier in the chapter. The two machine operators, both very young men, were noticeably nervous, but did a great job. The sales engineer who began the event as the naysayer was now a champion for lean. The external participants thanked the host company for allowing them to participate and learn, and offered encouragement for all to continue on the lean safety path. The two senior team members, the supervisor and lead man, were noticeably proud of what had been accomplished and said so when they presented. They recognized, along with the sales engineer and the business owners, that a culture shift had just occurred. A glimpse of the future had just been presented to them.

To add frosting to their cake, the team now led all 50 employees to the new work cell. The two young operators proudly took up their positions and ran a few parts through the process. It was recognized by everyone that real change had occurred and they all applauded when the demonstration was completed. The team was justifiably proud of what they had accomplished and gathered in front of the new work cell for a team photo before returning to the meeting room. I took the opportunity to thank all of them for individually contributing to the team's success. They had, in just three days, established strong relationships with each other because we had a common goal and accomplished great things together.

The following week I sent an e-mail to the president and vice president of the company thanking them for hosting the event and for allowing me and the loss control representative to add value to the trust by facilitating the event. I also asked them for a quote I could use in an article about the event I was planning for a monthly lean newsletter. They both responded and their responses support the theme of this book. Each of them in separate e-mails acknowledged that safety and lean had both taken a big leap forward because of this one event. Even more important to them was that they now had a team of people who were charged up and ready to continue implementing safety-focused lean in their plant. Within a few weeks, I heard from the president who informed me that there were two kaizen events going on in the plant. The former naysayer, the sales engineer, had taken on the facilitator role and was now leading the kaizen team. Needless to say, I felt good about this news and asked him to keep me informed about their progress. About a month later, I received an e-mail from their chief engineer, who was asked to be the champion for lean, asking if I would like to attend a kaizen event wrap-up meeting that the team was planning. Of course I said, "yes," and left work on the scheduled day to make their late afternoon meeting time.

Once again, the entire company had gathered in the cafeteria, along with a few guests including me, the loss control representative, and one of their customers. The five individuals from the plant who had participated in the first kaizen event made up this new team. After entering the building, I began encountering each of them and had a chance to say hello and wish them good luck during the presentation. Someone I worked with for years always said something like, "luck is when preparedness meets opportunity." I could tell these five individuals were prepared and they were about to be given the opportunity to shine. The meeting started with all five of them standing in front of their co-workers. They reviewed their team journey following the same agenda that was used during the first event. Noticeably less nervous this time, they described the current and future state conditions and many of the improvements that had been identified and implemented. Most noteworthy was the declaration that they had implemented 12 safety improvements. They took the time to thank all of their co-workers who had assisted them in their efforts and then invited everyone to join them in a tour of the newly configured workstations. The first stop was a large punch press. The team had eliminated some of the manual material handling of the parts by using a wide conveyor to transfer the parts from the press discharge area to the unloading area. The lead man, from our first event, provided a demonstration showing that the operator could now simply lift and position the parts onto a pallet while standing erect. Applying what they learned in the first event, they laid the parts horizontally on the pallet and had them correctly oriented for the next workstation operator. In the past, two operators would have been positioned at the press when running this part.

Because of the material handling safety improvements, only one was now required. Next stop on the tour was a work center where two secondary operations were completed. Here again, they had minimized injury risk by delivering parts to the operator at waist height, reduced the number of physical steps required to chamfer four holes on the top of the part, and eliminated any bending when placing the part into a shipping container. The two young operators from our first kaizen described the sequence of steps at this work center. I was amazed at how they had grown in confidence. When it ended, they were on cloud nine and so was I as proud as a parent.

What often occurs, when something really great happens and the people involved are all abuzz and do not want to leave, was occurring at this point in time. It is hard to imagine a group of individuals who were more excited about their success. I stated earlier that "the biggest benefit from a kaizen blitz is not the process you improve, it is the minds you change." Another favorite saying is, "Success is the greatest motivator." Attending this lean learning graduation ceremony for my kaizen blitz students reinforced both of these beliefs.

There are other potential business impacts resulting from this safety-focused kaizen activity. One is that this plant was becoming a "sales and marketing" tool. The fact that a customer had been invited to the presentation and tour was evidence that the sales engineer and the company leadership realized the value of lean safety to their business. Imagine a potential customer visiting and the tour being conducted by the two young machine operators who are passionate about lean and safety. In the future, maybe they will be adding sales commissions to their weekly pay. The other impact area is the obvious: safety. The cost of safety is not just measured in dollars, but also in the pain and suffering of those who are injured. There is no better way to quickly and effectively train a group of people to recognize and eliminate, or at least reduce, ergonomic-related soft tissue injury risks than conducting a safety kaizen blitz. If I have conveyed that message effectively with this case study, then I have justified whatever price you paid for this book. Whatever else you may glean and learn is frosting on the cake.

In the next chapter on safety program leadership, I will cover the value of using an employee-based safety team. Later, in Chapter 11 on metrics, I will review team-based safety metrics that can be used to encourage proactive "safety first" thinking and awareness. Teamwork works.

Quick Guide: Advanced Lean Tools for Safety

- A3: Use this lean tool to develop lean leaders who understand that managers are best when they manage processes and lead people. Use the A3 to redefine safety processes, such as the accident investigation process.
- Kaizen blitz: The kaizen blitz, or a rapid continuous improvement event, is the most powerful people-development and engagement tool in a lean thinker's toolbox. Use it to focus on safety improvement rather than cycle time gains.

Endnote

1. Wikipedia (2008) Ergonomics (overview): http://en.wikipedia.org/wiki/Ergonomics#cite_note-1

Chapter 6

Safety Program Leadership

It Doesn't Have to Be a Manager

How you use your existing staff and organizational structure to support and drive safety improvement is an important consideration when building world-class safety. Safety is a distributed responsibility in that everyone working for a company has some responsibility for safety. It may be for his/her own safety or the safety of others or even the entire organization. To begin to understand the current state of a company's safety program leadership structure and identify who is responsible for the varying parts of the program, a lean approach would be to identify all of the safety processes that comprise your safety program. Once that step is completed, those responsible can be aligned with each of the upper level safety processes. I would be willing to bet that you will identify some safety processes with unclear ownership. This usually is the case because safety is everyone's responsibility and we work in functional departments. Simply ask the question, "Who is responsible for safety in the engineering (or name your favorite) department?" The most likely answer to this question is the motherhood and apple pie response: "everyone." Safety is in every job description yet there are few job descriptions with safety in the job title. It is always farther down in the document near the catchall phrase: "and other duties as assigned." Yes, we are all indeed responsible for safety, but what world-class safety requires are people who are passionate about safety. Locating and engaging people with passion for safety is an early requirement in building world-class safety.

In Chapter 3, I referenced the role of safety directors and how, regardless of their inherent ability and passion, they cannot, by themselves, guarantee a safe workplace that is injury free. One of the other problems with having persons with "safety" in their job title is the avenue it opens for others to believe safety is just that—the safety director's job, and not theirs. No one person, CEO, COO, vice president of manufacturing, director of manufacturing, or anyone else, can guarantee a safe workplace. But, whoever is at the top of the organizational chart is ultimately responsible for safety in the business they are paid to lead. They may have delegated some of their responsibility down the organizational chart and have little or no involvement in the safety activities of the business. This chapter on safety leadership is not intended to define the best possible organizational chart to support world-class safety because all businesses and their organizational structures are different. Instead, this chapter is a deliberate attempt to help define a safety management structure that might be applied in any organization. By putting the right structure in place, it will help to identify people with a passion for safety who can build and manage a comprehensive world-class safety program.

Many companies default to their existing management structure to manage their safety program. If management is ultimately responsible for safety, this seems to make sense. This decision, though, implies a top-down approach to safety management, supported by most organizational charts that reflect a top-down approach to managing the business. The problem I see with this approach lies in the fact that a passion for safety is not evident in most supervisors. They have myriad tasks and responsibilities for which they are accountable. Safety is thrown into that mix, and possibly they will say, "Safety is my first priority," but in reality it can only be a part-time commitment. "Involvement in safety" is not the same as "passionate about safety" any more than "professionally managed" always means "well managed." Remember, we are seeking passionate people, so building a safety management structure that will identify these individuals will be better than using your existing management structure.

In the last chapter, one of the lean tools recommended for the building of world-class safety was teams. A formal team-based structure that distributes responsibility for safety process ownership is a great approach. Safety team membership requirements can be based on desire or interest rather than title. It can be cross functional because the team can and should be staffed from all areas and all levels of the organization. Let's start by identifying some of the key positions that would have to be filled on any safety team.

Team Leader

The beginning head of this chapter suggests that this individual does not have to be a manager. Here is why I believe that is the best approach. First, managers already have the opportunity to lead based on their position. Why not use this opportunity for leadership to grow some of your staff who show high potential. You never know how good someone is until you give him or her a chance to fail. Secondly, as you fill and then rotate people in and out of this safety leadership position, you build a core group of passionate safety experts. This safety team leadership scheme isn't an approach I just dreamed up, I have watched it work very effectively for many years. The personal growth I have observed in every individual who has taken on this leadership role is in alignment with the goal of lean: engaged people focused on business improvement. Each of them was forever changed. Their safety expertise and passion grew during their leadership tenure. All of them seemed reluctant to let go when their two- to three-year term had been completed. This was a sign that they were truly challenged and over time had developed a passion for safety that would move any company closer to world-class safety. All of these individuals had a real full-time position doing something else, yet they made safety their first priority. Everyone understands how important safety is, so anyone who is given this chance to lead and accepts the opportunity, will do a trustworthy job.

Management Facilitator

A senior-level manager should be responsible for safety in the facility. The safety team is his vehicle to drive world-class safety. The primary responsibility of this safety team member is that of a facilitator. Remember that facilitators, like midwives, assist, but do not deliver the results. Therefore, mentoring the team leader and clearing roadblocks for the team, rather than direct involvement in task completion, are appropriate activities.

Accident Investigation Facilitator

A future chapter will be devoted to effective incident/accident investigations. Every incident (near miss) or accident provides a continuous improvement opportunity that should result in root cause analysis and corrective actions. Here is a chance to utilize someone with a passion for lean, and his/her problem-solving skill sets, to fill this safety team position. Filling this role provides another opportunity to develop a future business leader.

Recorder Keeper

Documentation is a critical part of any safety program. Mandated safety logs, task lists, meeting minutes, program description documents, safety check sheets, and safety audit results are just some of the safety documents that have to be maintained. A key element of a world-class safety program is the maintenance of individual program description documents for each of the major elements of the safety program. For instance, to name just a few, the LOTO (lockout tagout) program, forklift training program, fire drill program, along with all other key elements of the safety program should be described in a separate document. This supports continual improvement because one can then schedule an annual review of each program document. The recorder keeper can meet with the subteam leaders, or owners, of all of the key programs to conduct the annual review. This step ensures the programs are updated as requirements change. It is the PDCA (plan-do-check-act) step in the continuous improvement of your world-class safety program. Seek out someone with strong organizational skills for this safety team role.

Subteam Leaders

Safety programs are a compilation of subprocesses or elements that all require leadership if they are to be effective. Subteams, with subteam leaders, are an approach that meshes nicely with this team approach to safety management. Each subteam leadership position you create opens up another opportunity to grow and develop future leaders. Here is a short list of some subteams that would help to get a company moving toward world-class safety.

■ *Lockout/tagout (LOTO)*: This is a critical part of every safety program. Adherence to the OSHA guidelines surrounding LOTO will help guarantee the prevention of serious injuries. Included in this subteam's responsibilities would be the creation and management of LOTO visual documents for each piece of equipment that has stored energy. Using the lean visual factory methodology would result in documents that contain digital photographs that clearly spell out the LOTO steps for anyone authorized to perform the LOTO procedure. An example of such a document is shown in Figure 6.1. This sample does not spell out all of the possible energy sources that could require isolation. It is only intended to represent the visual nature of LOTO documents in

Equipment ID	38765
Equipment name	Power punch press
Equipment location	Press department

Compressed Air Source		Electrical Source	
Device	Throw arm	Device	Ball valve
Location	On wall next to machine	Location	Left side rear of machine
De-energize	Pull arm down. Lock and tag	De-energize	Turn handle 90 degrees, cap, lock and tag
Verify	Press power on button on machine main control	Verify	Test air hose
Approved by		**Date**	

Figure 6.1   A sample of a lockout/tagout (LOTO) form is shown.

a lean environment. Defining a process to ensure document updates occur as new equipment is installed, or equipment is removed, is also required. Conducting an annual audit of the LOTO program would be another responsibility of this subteam. They should select around 10 work centers, or pieces of equipment, and then ask the operators who normally operate that equipment to perform the LOTO procedure while they observe. To assess the LOTO training program effectiveness, they should select a few of the newest operators to be included in this audit.

■ *Personal protective equipment* (*PPE*): This subteam's responsibility is to ensure the appropriate PPE has been defined for all work tasks in the plant and that the information has been communicated to all with a need to know. The lean visual factory approach is again the way to go. Visual documents that display the PPE requirements by department, work cell, or individual piece of equipment may be appropriate. Visuals of the required PPE, such as safety shoes, safety glasses with side shields, gloves, ear plugs, etc., can be pictured on the posted documents. Few words are required to convey the requirement to all when a lean visual system is put into place.

■ *Ergonomics*: If soft tissue or musculoskeletal disorders (MSDs) are a safety concern or an ongoing problem, a proactive way to address them is to form a subteam charged with training and educating the workforce in the basic methods to reduce these types of injuries. At work centers or on jobs where these types of injuries have occurred, this subteam may marshal a safety kaizen blitz team to evaluate and reduce the risks of future MSD injuries.

■ *Hazardous material (HAZMAT)*: Effective HMIS (hazardous material identification system) and MSDS (material safety data sheet) programs are both essential elements of a safety program. The HMIS system identifies and labels substances that may pose some hazard to those who use them. The MSD sheets, provided by the suppliers of materials deemed to be hazardous, are kept on file to quickly identify all hazards and the appropriate responses in emergency situations. This mandated part of a safety program can be handed off to a subteam that ensures the MSDS database in updated and the HMIS labeling program is audited for compliance.

■ *Safety promotion*: To help engage everyone in thinking safety, most safety programs utilize games and/or rewards to help raise safety awareness. In Chapter 8 on promoting safety, overviews of proactive safety improvement initiatives will be covered in some detail. For now, it is important to recognize both the need for this element of a world-class safety program and someone to lead the effort.

This use of subteams to take ownership of portions of your safety program can be expanded or contracted as needs change. This is not a complete list of the possibilities, but is offered as a representative example of the types of subteams that can be formed.

Human Resources (HR) Facilitator

The responsibility for maintaining OSHA-required logs and other records generally falls upon your HR function if you do not have a full-time safety person. This is a natural and appropriate assignment since the HR function also maintains the workers compensation and medical records linked to a company's safety program incidents and accidents. Another benefit of having an HR representative on your safety team is his/her ability to facilitate and develop individuals on the team. Most everyone who works in an HR role possesses strong interpersonal skills, which they bring to every meeting. They can be a real asset to the team.

Safety Walk Co-Coordinator

Safety walks, or informal audits, are almost always an element of safety programs. An individual or individuals use some form of a checklist as they walk through a facility looking for safety violations. Based on my experience, the forms usually focus the attention of the auditors on things, such as equipment, rather than on people's actions. Most injuries that occur are a result of an individual's actions. A world-class safety program would include this audit step and have an audit process that requires the auditors to observe work activities as well as the equipment. The safety team owner of this audit process may elect to engage others in the auditing. One method of doing this is to have assigned safety contacts in each work cell or department. When the audit is conducted, this safety contact would join in the audit. Any findings that require corrective action can be immediately addressed by this individual in coordination with the area supervisor.

OSHA Knowledge Expert

The person who fills this job title will most likely be from your manufacturing engineering staff. Knowledge of and compliance with OSHA regulations requires a broad body of safety knowledge. If a resource like that is unavailable, a business can take advantage of externally offered training classes or bring in consultants to assess compliance. Another opportunity for small businesses that may lack resources is to partner with OSHA. The following overview was taken from the OSHA Web site:

The OSHA Strategic Partnership Program (OSPP) for Worker Safety and Health, adopted on November 13, 1998, is an expansion and formalization of OSHA's substantial experience with voluntary programs.

■ In a Partnership, OSHA enters into an extended, voluntary, cooperative relationship with groups of employers, employees, and employee representatives (sometimes including other stakeholders, and sometimes involving only one employer) in order to encourage, assist, and recognize their efforts to eliminate serious hazards and achieve a high level of worker safety and health.

■ Partnering with OSHA is appropriate for the many employers who want to do the right thing, but need help in strengthening worker safety and health at their worksites. Within the OSPP, management, labor, and OSHA are proving that old adversaries can become new allies committed to cooperative solutions to the problems of worker safety and health.

■ OSHA and its partners can identify a common goal, develop plans for achieving that goal, and cooperate in implementation.

■ OSHA's interest in cooperative Partnerships in no way reduces its ongoing commitment to enforcing the requirements of the Occupational Safety and Health Act. The OSPP moves away from traditional enforcement methods that target individual worksites and punish employers who violate agency standards. Instead, in a growing number of local and national Partnerships, OSHA is working cooperatively with groups of employers and workers to identify the most serious workplace hazards, develop workplace-appropriate safety and health management systems, share resources, and find effective ways to reduce worker injuries, illnesses, and deaths.

■ Most of the worksites that have chosen to partner with OSHA are small businesses, with an average employment of fewer than 50 employees.[1]

Training Coordinator

Mandated training requirements related to LOTO, forklift operations, and HAZMAT, just to name a few, must be provided and training records maintained. Having someone on the team who is responsible for this training is a way to ensure these compliance standards are met. In addition, many other safety training opportunities exist. The team training coordinator may recommend fire safety, hearing protection, proper lifting techniques, and many other training opportunities.

If all of the above listed positions are filled, you will have a large safety committee. The advantage of this is that many safety champions are being developed at one time. That is in alignment with lean thinking—engaging employees in continual improvement activity. Smaller companies may not have the resources to staff a team of that size and may have to ask participants to wear more than one hat.

One essential element of a team-based safety program is a team task list. The safety team's task list is the company's safety program outline document. It guides the team by listing open tasks, annual reviews, and training requirements, successes, and other relevant information. It should be used by the team leader to conduct a bi-weekly meeting of the safety team. Their agenda could be as follows:

1. Review any open action items from the current task list and add new tasks.
   a. Assign responsibility for any tasks added to the list during the meeting.
   b. Ensure a target date is set for each task.
2. Review any open incident/accident investigations.
   a. Open incident/accident investigations should be tracked. Columns that list target dates for completion and percentage complete for each line item will stress the importance of completing the tasks assigned during the investigation process.
3. Subteam updates, e.g., annual reviews, audits results, safety promotion activities, upcoming training, etc.
   a. This portion of the meeting gives the subteam leaders the opportunity to update the team on any activities within their subteams.
4. Safety walk results.
   a. Review any open tasks that resulted from plant safety walks.
5. Metrics.
   a. A review of mandated and internally defined performance measurements is critical. In order for the safety team to feel like a team, they require a common focus—something that pulls all of their varying activities together. Reviewing the metrics and targets they helped to define serves that purpose.
6. Successes.
   a. It is very important to always recognize any and all recent successes. Everyone on the team should have a chance to recognize events

or milestones that they consider a success for the safety program. Collect these items at the bottom on the team task list and add to the list as the year passes.

7. Open discussion.

    a. As members of the safety team, they will be approached by others with safety questions that they cannot always answer. This is the time and place for open discussion on those questions or topics that may relate to safety policies.

A well-defined and maintained task list simplifies the team leader's job. Without a well-defined and maintained task list, the meeting breaks down into a safety discussion that often produces little in the way of results. Good leadership and facilitation support should limit the duration of safety meetings to about one hour. Sticking to the bi-weekly schedule is essential because canceling a safety meeting sends the wrong message about safety.

Obviously, leadership of safety must be provided from many different levels of a business. Coaching, mentoring, and facilitating a safety team may be more work than just managing safety yourself. Letting go, and relying on a safety team to manage a safety program, is a gift management can give to others so that they can grow as a result of new and challenging responsibilities. It is extending trust to earn trust. Lean leaders understand that growing a business is all about growing people. Don't pass up this opportunity to build a team of passionate safety leaders in your business.

Quick Guide: Safety Program Leadership

■ Locating and engaging people with passion for safety is an early requirement in building world-class safety.

■ No one person, CEO, COO, vice president of manufacturing, director of manufacturing, or anyone else, can guarantee a safe workplace.

■ "Involvement in safety" is not the same as "passionate about safety."

■ A formal team-based structure that distributes responsibility for safety process ownership is a great approach.

■ Use this opportunity for safety team leadership to grow some of your high potential people.

- Management facilitator—Mentoring the team leader and clearing road-blocks for the team, rather than direct involvement in task completion, are appropriate activities.
- Safety subteam leadership positions open up another opportunity to grow and develop future leaders.
- An HR role provides interpersonal skills and OSHA documentation expertise.

Endnote

1. United States Department of Labor (2006) Strategic partnership overview: http://www.osha.gov/dcsp/partnerships/what_is.html

Chapter 7

Incident/Accident Investigation

Getting to Root Causes and Lasting Solutions

No one comes to work with the intention of getting injured. Incidents and injuries both repetitively occur and are almost always treated the same, as negative occurrences. Much of this negative thinking is driven by the fact that most mandated safety metrics track negatives. Clearly, anytime someone is injured, it absolutely is detrimental because they have endured pain and suffering. Yet, lean thinkers understand incidents and accidents are not exactly the same and should be approached differently. Synonyms for accident include words, such as *calamity, catastrophe, misfortune,* and *mishap.* The definition for accident includes phrases like "the way things happened without any planning," and "an unplanned and unfortunate event." It almost sounds as though accidents are just fate and nothing can be done to prevent them. Synonyms for incident include words, such as *event, occurrence,* and *occasion,* and the definition includes wording like: "something that happens" and "an event that may result in a crisis." To lean thinkers, incident is a word filled with opportunity. When a serious accident happens, reactions are the response. Red-faced managers lead the accident investigation and make the "safety first" pledge once again. When incidents occur, and are investigated with a lean mindset and tools, proactive actions can come about to prevent future accidents. Incidents are not negatives, but instead a goldmine of safety improvement opportunities waiting to be mined. Lean thinkers see occurrences, such as safety incidents, as their bread and butter, while others see them as problems. Lean thinkers would all be unemployed if it weren't for business problems. Their skill set can be used to turn problems like safety incidents into safety improvements. Incident investigations are yet another opportunity for lean thinkers to engage the workforce in a facilitated safety improvement activity.

Focus on Process, Not People

The reason a lean thinker can turn, what seems to most, a negative safety occurrence into a positive is because they will focus everyone on the process, not the person. What and why the incident occurred have to be identified to prevent the reoccurrence, not who did it. Too often, in years past, the result of accident investigations was discipline being issued to the individual involved. This does not build trust, but widens the trust gap. When I was much younger, I was involved in a workplace incident. I had recently returned to a factory job at a large steel plant after serving four years in the military. I was a machine operator running equipment that produced barbed wire. It was difficult, hard work that entailed running three machines that each produced a large spool of finished barbed wire every 12 minutes. When a spool was completed, the operator, while bent over and using wire cutting pliers, had to cut the line wire, drop the spool, drag it out from under the machine, secure the lose end, and then, using smaller gauge wires, secure the spool to a wire frame onto which it had been wound. The spool and the operator were standing on oily ½-inch-thick steel plates that were in between the machines. The pallet onto which the spools had to be stacked was at the rear of the machines. To get the individual spools back to the pallet, the accepted practice was to use a steel bar, with a hook at one end, to push or pull the spool of wire a distance of about 12 feet on the steel plates toward the pallet. The steel plates ended about five feet before the pallet and the floor was constructed out of wooden blocks. When the spool hit the wooden block, the barbs dug in and it stopped. Most operators, including me, pulled the spool and then stepped to one side while pulling it hard enough so it would continue sliding and then bounce over the wooden floor to the pallet. A hoist was then used to place the spools onto the pallet. At the end of one of my shifts, my supervisor handed me a "safety observation" form that noted I had violated safety policy by pulling rather than pushing the spools.

Safety observations were part of the company safety program. Managers had a bogey of two per month and were discouraged from writing up positive observations (observing someone adhering to safety policy). It was believed by management above the supervisory level that observing and writing up unsafe acts, via these safety observation forms, would have a positive impact on plant safety. The supervisors, needless to say, did not like to write up unsafe acts and preferred instead to write up positive safety observations. So, there I stood holding the piece of paper while the supervisor walked hastily away. He had done the dirty deed and was half way to his monthly target of two observations. That is the end of the story except for the fact that after all these years I still remember the situation. Management was right in their thinking that issuing safety observations for unsafe acts would impact one's thinking. I remember thinking that everyone else in the department often pulled rather than pushed the spools and that I had been singled out. It was not a positive experience for me and it widened the trust gap. Then, a few weeks later, while stepping over the spools I had pushed back toward the pallet, my shin hit the top edge of a spool and I ended up in the clinic to have a nasty gash cleaned up that needed stitches. Obviously, the act of either pushing or pulling the spools had some inherent danger present. Looking back, wearing my "lean" lenses, a few glaring problems (that means opportunities to me) existed in this situation.

1. Supervision did not understand that being honest with people would allow them to grow.
2. The responsibility to engage employees in discussions to find ways to eliminate or lessen the safety hazards that existed in this work process was management's.
3. A belief that discipline (the safety observation) was a way to motivate individuals and improve safety behaviors.
4. There was no systematic process for safety improvement.

Obviously, that was many years ago, but I would like to be able to go back in time and facilitate a kaizen blitz event to reduce the risks associated with that job. I am sure the kaizen team would have experimented with new layouts that would position the pallet closer to the spool discharge point, small roller dollies or hoists to transport the spools, protective shin guards, and many other ideas to make the task both safer, easier, and, therefore, faster. One thing I understand for sure is that I cannot change the past, but we can all learn from it. Let's contrast that safety incident situation with one that is more time relevant.

Getting to Root Cause and Corrective Actions

I had begun to investigate every accident that was reported in the facility where I worked after I visited another plant. As often happens in the lean community, we steal ideas with reckless abandon from each other. It is an accepted and expected activity. So, while I was visiting this other plant and talking with the plant manager to seek approval for a possible tour of their facility, I noticed a stack of safety accident forms on a small table in front of me. When I asked her about the forms, she stated that anyone who is injured on the job has to come into her office and complete the form with her. I honestly don't know if this was to intimidate the employees or if it was a sincere effort to get to root causes and corrective actions. I would like to think the second reason is the correct answer. My takeaway was that she was involved in every investigation, no hand-offs; safety was first. I decided to start leading the investigations for any injury that occurred in our facility to get to root causes and corrective actions. Then at the end of that year, when the safety incidents and accidents were totaled and put into a Pareto chart (a chart where values being plotted are arranged in descending order), we recognized that forklift incidents were a major contributor to the total. A few managers investigated force monitoring devices that would kill the forklift ignition system when the forklift made contact with anything with a force above the predetermined setting. A supervisor would then have to be called to restart the forklift. I suggested they not do that for the obvious reasons (you destroy trust when you police people) and offered instead to lead the investigation of every forklift incident to see if we could drive down the incident rate.

When we started, most of the forklift incidents that were reported were documented by the person who found the damage and we had no idea who was driving the forklift when the damage occurred. They were hit and run accidents. Because of that, when the investigations were scheduled, we invited a forklift driver from the work area or the shift to be part of the investigating team. I still remember their looks of concern and state of nervousness when they walked into my office. They, of course, were thinking that we thought they might have been responsible for the accident and resulting damage. Their concern began to abate when I informed them that we would not be spending any time trying to find out who was driving the forklift. Our goal was to determine the root cause and then put some corrective actions into place that would prevent a reoccurrence. That is exactly what we did for every forklift incident. Root causes of many or most of the forklift incidents had to do with a lack of clearance. We were asking forklift drivers to be so precise that it was almost impossible for them do their job without coming into contact with something. In some cases, we widened the aisles where they had to work and, in others, racks were moved to allow them easier access to storage locations. Equipment, signs, fire pull switches, and other items were moved out of their work paths so they could drive without hitting them. It didn't take long for the forklift drivers to start completing the forms themselves when they had an incident. They no longer feared discipline. They gladly came into my office to participate in the continuous improvement of safety. We had earned their trust by focusing on the process of driving a forklift not on the forklift driver. I wish I could have had that same opportunity, to participate in safety improvement, many years ago in the steel plant. My supervisor would have earned my trust if he had engaged me in a positive experience rather than handing me a form and walking away. Don't pass up these golden opportunities in your plant. Incident and accident investigations are perfect forums in which to engage the workforce in the continual improvement of safety. Lean tools, like root cause analysis or "asking 'why' five times," used by a lean thinking facilitator can turn every incident into a safety improvement and every accident investigation into the best possible outcome. Even though someone already suffered an injury, you can still define corrective actions to prevent a reoccurrence.

At times, when two people are involved in a safety incident or accident, emotions may be running high. Accusations of unsafe acts may be hurled around and you may find yourself acting as a mediator rather than a facilitator. When I think this situation could occur in an investigation, I take time to select the correct lean tool before the investigation begins. One I often rely on is the future state process map. My reasoning is that I want to delay the possible emotional outbursts related to the incident until I have both parties consider the best possible and safest process. I simply start the meeting by noting that we are here to investigate an incident and there may be some emotion surrounding the situation that occurred, so before we talk about what happened, let's together map out the steps of the safest possible process. This activity forces them to reflect on the safety process relative to their own actions during the incident. If they were accountable for an unsafe action, the future state map will point that out to them without anyone having to say it. By the time the future state "safest process" has been mapped and they have had a chance to reflect on their actions during the incident, they have calmed down. Any discussions after that point, about who did or didn't do something, can be compared to the "safest process" they helped to define. This reversal of the standard process mapping steps, current state, and then future state, works very effectively to defuse emotion.

Here is another opportunity to use the process mapping lean tool. One should process map the incident/accident reporting, investigation, and tracking process steps so that everyone in the business clearly understands their duties. This very important step in rebuilding a safety program should be the responsibility of those who own the distributed processes. Included could be operation's management, HR representatives, frontline supervisors, employee safety committee representatives, the safety manager (if a firm is large enough to staff that position), and a lean facilitator. Start the current state mapping exercise with a fictitious incident or accident and record all of the steps required until the incident or accident case is closed. In many businesses, this exercise has never taken place. As with most distributed processes, people do their individual parts, but no one ever looks at the whole. When completed, the map will define the current state of incident/accident reporting, investigation, and tracking, and will be the baseline for continuous improvement of this important business process. Illustrated in Figure 7.1 is an example map of an incident reporting and investigation process.

When conducting the accident or incident investigation, try and use a neutral room rather than a manager's office. That will help set people at ease. Included in the investigation meeting should be the individual or individuals involved in the incident, their supervisor, an HR representative, and a facilitator to lead the investigation. Others can be added as appropriate. For instance, if someone is responsible for all forklift operations in a facility, they should be invited to all incidents involving a forklift. Following lean philosophy, get the investigation team out on the shop floor at the site of the incident rather than conduct the entire investigation in the meeting room. Facts are more important than what people think happened and the only way to get the facts is by going to the scene of the incident. I am often surprised how many misconceptions are cleared up by following this important step in the investigation process.

If someone is injured at work, the last statement I make during the accident investigations is: "I am sorry you were hurt at work." I say it like I mean it because I do. Even when they have violated known safety rules, I say it. Even if they are disciplined for that safety violation, I say it. It is important

Incident Initial Response and Investigation Process

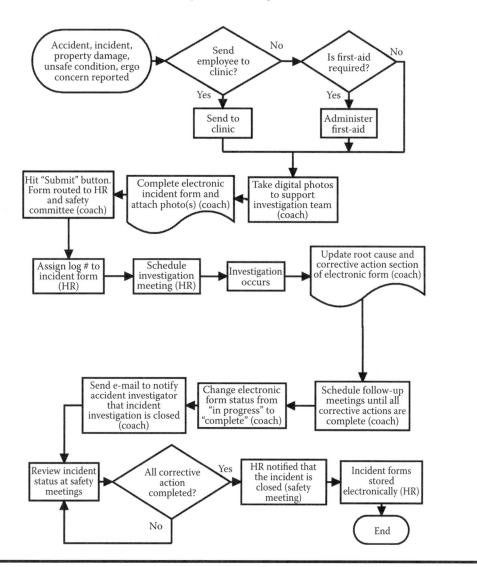

Figure 7.1   Process map of incident reporting and investigations.

to say it because people do not come to work with the intention of getting injured or injuring themselves. Remember, people do not care how much you know about safety until they know how much you care about their safety.

This brief overview on the usefulness of lean thinking and lean tools when investigating accidents certainly does not fully explore the topic of accident investigations. There are many excellent resources on the Internet, including the OSHA site (www.OSHA.gov), which give detailed instructions. All of this information is free for the taking and can be used in conjunction with the lean tools mentioned to build a world-class process that will help to drive down your accident and incident rates. Engaging your employees in accident/incident investigations to get to root causes and corrective actions builds trust. I have witnessed the power of this experience over and over again.

Quick Guide: Lean Approach to Incident and Accident Investigations

- All incident or accident investigations are situational.
- Focus on the what and why and not the who during investigations.
- Get to root cause(s) and corrective actions to prevent reoccurrences.
- Treat every incident as an opportunity to improve safety.
- Recognize that most often the process, and not the person, is the problem.
- Drive fear out of the investigation process.
- Earn the trust of those involved.
- Be honest with those who have violated safety rules or practices.
- Go to the *gemba* (shop floor) to see and gather facts.
- Focus on the underlying work process.
- Use lean tools, 5-whys, and process maps, to get to root causes.
- Define owners and target dates for all identified tasks.
- Schedule follow-up meeting to ensure task completion.

Chapter 8

Promoting Safety

Engaging Employees to Build Safety Awareness

Building safety awareness is the goal of every company with a safety program. Awareness is considered the key to a good safety program because it assumes an individual will think "safety first" before taking any action. Consider the safety first analogy of a pedestrian looking left and right before crossing a busy street. It is an ingrained safety principle we all adhere to because it was drilled into our heads from the time we learned to walk. With the exception of a few American tourists in London, who only look left and then step into oncoming traffic in the right lane, pedestrian-related accidents are low. How can a company raise safety awareness to that level in their plant? Not by playing safety bingo. That had been one of our safety awareness programs many years ago. Each day our receptionist would spin the cage containing the little wooden balls and announce the number over the public address system. Many employees, statistically probably the same number who take pleasure in gambling, watched their cards religiously as the receptionist announced the "safety bingo" number each day. After a while, the receptionist no longer even used the word "safety" when announcing the number. It had become the "bingo number for today" announcement. Without using the word "safety," she eliminated the one and only chance there was to raise safety awareness. Safety bingo, and any other game like it, does nothing to raise safety consciousness. A much more effective method of raising safety awareness using the paging system is to have someone announce each and every day the number of days since the last "lost time" accident. Most companies track that metric and often post it near the front entrance of their business. People walk right by it without taking notice because it is part of the everyday landscape. Using the paging system gives it focus and can raise the safety awareness of everyone who hears the announcement. Here are some other methods that are guaranteed to raise safety awareness because each of them engages individuals in safety-related activities.

Flash Meetings

A common practice in companies on the lean journey is the morning (or start of the shift) flash meetings in each work cell or department. They are intended to be 10-minute meetings with an agenda that deals with current issues. Topics can include: who is and who is not there, a review of the work load, equipment staffing assignments, and other operational issues. These meetings are a perfect forum to discuss safety at the start of every day or shift. To engage others in safety talk, start the meeting by asking if anyone has any safety concerns. Providing an open forum for safety feedback and follow-up actions sends the message that safety is important. Other safety topics could include the review of a weekly safety focus. Simply take an annual calendar and identify, with your safety committee, a safety theme for each month and a related topic for each week. Fire safety could be a focus during the month in which you conduct your annual fire drill. Figure 8.1 shows a list of monthly safety themes for the first four months of the year.

Another safety topic could be a review of safety metrics. The number of days since the last accident or incident could be a topic. Or consider combining a lean visual factory tool with the team's safety metric. During the flash meeting, have someone on the team affix a green cross safety sticker onto a monthly calendar each morning, for the previous day, if there were no injuries or incidents. A sample calendar that could be used with individuals was shown in Chapter 4, Figure 4.3. Starting each and every day with "safety first" topics raises safety awareness.

Benchmarking (A Standard against Which Something Can Be Measured or Assessed)

A common practice for lean thinkers is to make benchmarking visits to other companies in order to "steal ruthlessly" any ideas that can be taken back and implemented in their facility. Benchmarking can be a very formal (using

Monthly Safety Focus List

	January LOTO	February Back Safety	March Hand Safety	April MSDS/HMIS
Week 1	Discuss location of safety lock, tags and procedures	Demonstrate correct lifting	Inspect tools for damage or worn surfaces	Review an HMIS label on a product in your cell
Week 2	Review a lockout procedure	Demonstrate how to correctly carry something and set it down	Wear protective gloves when using solvents	Demonstrate how to get an MSDS from the PC
Week 3	Perform a lockout procedure on the equipment with the most energy sources in your cell	Identify the items in your cell that are too heavy for one person to lift	Do you need gloves for the job? If so, are they the correct fit for the task? (ergonomically)	Discuss how to get an MSDS or HMIS info and label for a new product
Week 4	Audit all equipment in your cell for a lockout procedure	Discuss the mechanical lifting devices in your cell and inspect them	Think about where to put your hands…and where *not* to put your hands while working with tools/machinery	Demonstrate how you get a HMIS label for a secondary container

Figure 8.1   A monthly safety focus list is illustrated.

standard forms and predetermined steps) or a relatively informal process. The size of a business and the resources available will probably determine the formality level. As noted in Chapter 5, the company where I worked was a member of a self-funded workers compensation insurance trust. With around 100 member companies, we were a member of a safety community. Benchmarking visits were a common practice between members. It was a safety consortium. To build safety awareness, invite hourly members of your safety team to go with you on benchmarking visits. Give them an assignment to identify one or two safety practices that can be brought back and implemented. Another opportunity to find benchmark sites is by attending regional National Safety Council, or other safety organization, events. Once again, send hourly employees who usually do not have the opportunity. They will return with heightened safety awareness and, hopefully, some contact names for planning future benchmarking visits. And don't forget the greatest benchmarking opportunity of all—the World Wide Web. You can research and find information on any safety-related topic on the Internet. In the 1970s, there was a soft cover, large format book titled *Whole Earth Catalog*. The subtitle was: *Access to Tools*. The 1990s version is the Internet. It is much faster and gives you access to every safety tool you need to develop a world-class safety program.

Safety Improvement Programs

In our team-based culture, each factory team maintained six metrics by which they and management could judge their performance. When team-defined metrics were first introduced, the safety measurement all teams tracked was injuries. They set a goal of zero injuries for the year. Each quarter, during the team's continuous improvement meeting, all of their metrics and goals, including safety, were reviewed. At the end of the year, most team's safety scoreboards reflected the fact that they had zero injuries. I allowed them to feel good about that for a few minutes before informing them that a metric that provides no data is a worthless metric. I continued by informing them that the only reason we measure anything is to identify the problems. Problems, that if solved, would improve the process. Their metric told us nothing other than they had avoided the negatives: injuries. I suggested instead that they begin to measure something positive and proactive that might drive safety awareness. Within a short time, all of the teams began to measure safety improvements. Each team would target a number of improvements for the quarter. Then, in their quarterly review meetings, they would report on their performance against target. In the spirit of continuous improvement, if they attained the quarterly target, they were challenged to raise the bar for the next review period. The employee safety team reviewed all submitted safety improvements. A "safety dollar" was given to the originator of all approved safety improvements. These safety dollars could be accumulated and used to purchase safety-related items that were imprinted with the company logo. This was, and still is, an employee engagement method that raises safety awareness. It was common to have over 300 approved safety improvements in a year. The idea for a safety improvement program stemmed from our IPI (implemented process improvement) program. This was our employee engagement method used to collect and implement our employees' process improvement ideas. It was not a suggestion program because the originator of the idea had to implement the changes before submitting the IPI. Then, when our ISO (International Organization of Standardization) auditor gave us feedback saying that we did a great job of reacting to quality issues, but we lacked a proactive method to improve quality, we rolled out a quality improvement program that mirrored the IPI and safety improvement programs. All three of these programs engaged all of our employees in business process improvement. Anyone can engage his/her employees in safety improvement identification in order to drive safety awareness by benchmarking this method. Figure 8.2 is an example of a safety improvement program form.

Safety Observation Program

Another program that is very similar to the safety improvement program is a safety observation program. Earlier I mentioned that most accidents occur because of someone's actions. This program focuses on actions related to safety. Simply create a form that any employee can use to recognize a co-worker for a safe act. For instance, if someone is lifting a container and bends at the knees and keeps his back straight while lifting, someone could recognize him for this safe behavior. I still remember one I received. I had been walking down one of the main aisles of the plant when I noticed a metal part on the floor. I stopped, bent down, and picked it up. A few hours later, one of the machine operators, from that area, stopped by my office and handed me a safety observation form. On it he had noted my picking up a tripping hazard from the aisle. I felt really good until he informed me that my back could have been straighter when I bend to pick it up. Honest

Implemented Safety Improvement

Name: _____   Clock # _____

Department _____   Date completed _____

Describe your safety improvement

_____

_____

_____

_____

_____

_____

_____

What type of injury risk does it reduce?

_____

_____

_____

_____

_____

_____

_____

_____

Safety team approval Yes __ No __ Date approved _____

Comments

_____

_____

_____

_____

_____

_____

_____

Figure 8.2   An example of a safety improvement program form.

OBSERVE SAFE WORK HABITS

Safety Observation Coupon	Safety Observation Coupon
To: _____	To: _____
From: _____	From: _____
Describe the observed behavior that demonstrated "Safety First."	Describe the observed behavior that demonstrated "Safety First."
_____	_____
_____	_____
_____	_____
_____	_____
_____	_____
Give this copy to the person observed	**Give this copy to the Safety Team**

Figure 8.3   Sample of a safety observation program form.

safety feedback given in an environment of trust. Figure 8.3 is a sample form for a safety observation program.

Job Safety Analysis

Job Safety Analysis (JSA) is a formal method that can be used to identify, analyze, and record the steps of a job in order to identify the safety hazards and avoidance methods. Although this topic was already addressed in Chapter 4, it is important to restate the value of this tool here. JSAs are the perfect employee engagement tool with which to build safety awareness. Using a prescribed method and standard forms, work team members evaluate the very work they perform to identify any safety hazards. It is a safety kaizen activity.

Safety Kaizen Blitz Events

A good portion of Chapter 4 was devoted to the use of this lean tool to drive safety improvement via employee engagement. I have only listed it here because it is the king of safety promotion and employee engagement tools. It will give your safety efforts high visibility and senior management involvement when they attend the kaizen wrap-up meeting.

Engaging employees in safety improvement activities is the only way to raise safety awareness to world-class levels. All of the methods mentioned in this chapter will accomplish that, but only through the repetition of these activities will you get safety awareness to a level comparable to that of people looking both ways before crossing a street. Covering "safety first" at every flash meeting every day, numerous benchmarking visits, an ongoing multiyear safety improvement program, safety observation program, hundreds of JSAs, and multiple safety kaizen events every month will move you closer to the utopian goal of a fully engaged workforce that thinks "safety first" before every action. The connection between lean and safety, and the benefits of using them in parallel should be crystal clear.

Quick Guide: Lean Approach to Internal Safety Promotion

- Get as many employees as possible involved in safety improvement.
- Hold daily work cell flash meetings at which safety is an everyday topic.
  - Address any safety concerns.
  - Update and discuss team safety metrics.
- Benchmark other's programs.
  - Visit local companies.
  - Utilize the World Wide Web.
- Start a formal safety improvement program.
  - Establish process and a form to gather everyone's improvement ideas.
- Begin a formal safety observation program.
  - Develop process and a form to allow all to recognize other's safe behaviors.
- Begin a JSA (job safety analysis) program.
  - Develop program guidelines and forms. Begin with a pilot area.
- Hold safety focused kaizen blitz events.
  - Develop safety champions via this hands-on training method.

Chapter 9

Roadmap to World-Class Safety

Pulling the Pieces Together to Build or Rebuild a Safety Program

To assess a safety program, a starting point could be to list all of the elements of the existing program. This will provide an overview representation of the "current state." The next step (to benchmark a complete safety program) can be as simple as going online and purchasing, from any of a number of providers, a complete safety program outline. Now, in possession of your current state and a possible future state, the two can be compared to discover any gaps. The big gap in this process of safety program assessment is that the purchased program outline will tell anyone what he/she has to do, primarily compliance-related tasks, but the difficult part of defining how to do it is still up to the individual. Sure, the outline will suggest having a safety team, but how do you develop an engaged safety team that will take proactive ownership of safety? The answer is still the same: By impacting your culture using lean leadership and lean tools. Attaining world-class safety is not just about the elements of the safety program any more than real lean is only about the lean tools. Your ability to impact people and, therefore, the cultural side of safety and lean will determine your success. In each of the following program elements, I will suggest a lean thinking approach to assist with the defining of "how" to successfully implement each element.

Safety Program Leadership

As stated more than once already, safety programs are often broken into pieces and the leadership, or ownership, of the parts is distributed. Yet, someone, and they know who they are, is ultimately responsible for the safety of all of the company's employees. These are the leaders who state, "Safety is our no. 1 priority" and "Our employees are our most valuable assets." To have a world-class safety program, saying it is no longer enough.

Lean Approach to Safety Program Leadership

Leadership must be lean savvy. They have to understand the lean approaches of focusing on process and continual improvement via employee engagement. A safety policy statement must be written and provide leadership the option to include statements about "management involvement" rather than the typical "management commitment" statements. Another prerequisite is a thorough understanding of change management and the ability to lead change by means of actual hands-on involvement. Leadership needs to ensure the availability of the required resources: people, money, and time, for those entrusted and challenged with taking safety to a new level. Leaders must demonstrate that they are not just responsible for the financial results of the business, but also for the safety of every employee. Every manager also owns this last responsibility in the business. Each of them should develop some safety-related "standard work" activities that will publically demonstrate their safety commitment. These activities could range from simply walking the plant every week to review the posted safety metrics on team boards to the facilitation of a kaizen or job safety analysis (JSA) development team. Other opportunities include instead of having lunch with the vice president of marketing, take the safety team to lunch, or reviewing the list of approved safety improvements and writing a note to thank some of the submitters. I could go on and on and provide many more examples, but I think the message is crystal clear. If you really believe safety is the first priority, take some time out of your busy schedule to demonstrate it.

Safety Team

Groups of people who are responsible for safety are often referred to as a safety committee in conventional safety programs. The definition for committee is: "a group of people appointed or chosen to perform a function on behalf of a larger group." Contrast that with the definition for team: "a number of people organized to function cooperatively as a group." All safety committee members attend safety meetings while some tend to do most of the performing. Because committee membership is often based on job title and current responsibilities, the safety work tends to get distributed along those same organizational lines. It is just a way to get the work of managing safety done using the existing organizational structure and the people who fill those positions. Leadership of the committee probably falls to someone in an existing leadership position.

Lean Approach to a Safety Team

Teamwork, at some level, is a lean tool used by almost every company who has recently won the annual Industry Week magazine's "Top 10 Plants" competition. Any company that is serious about lean knows that a reliable proven path to engaged employees is the building of work teams. Empowered natural work teams, working cooperatively in customer-focused cells, will get a company to the lean end goal of reducing the time from customer order to shipment by eliminating waste quicker than any other method. Consider the advantage of having a safety team working cooperatively toward world-class safety, if that is your goal. Leadership of the team can come from any area of the company. Pick someone with ambition, ability, and an attitude to succeed that is not currently in a management role. Leading a safety team is a great way to showcase and develop the future talent within any business. Yes, they will require mentoring and coaching to be successful, but those are required elements of any leadership development program. A safety team leader who is a peer of those doing the work, the hourly employees, will automatically garner some respect as their spokesperson.

Selecting the rest of the team can still be influenced by the work that has to be accomplished. For instance, a human resources (HR) representative is mandatory because of mandated OSHA injury reporting and recordkeeping requirements. Once the major elements of a safety program are clear, the team positions required to manage these elements will surface. To fill the positions, you can ask individuals that you would like to develop or you can develop job descriptions and post the positions for the safety team openings. Either way, a cross-functional team that broadly represents the organization ensures that your safety program isn't just a shop floor program. Figure 9.1 is a job description for the safety contact position. Job descriptions for some or all of the safety committee roles, such as the many subteams listed in the

Safety Team Job Descriptions

Department Safety Contact

- Understands the company's safety policies and program
- Leads discussion on safety topics during team's morning flash meeting
- Walks new departmental employees through the safety check sheet (PPE, evacuation procedures, LOTO, HMIS, first-aid, etc.)
- Acts as a safety-buddy for new employees to ensure they have someone they are comfortable to go to on any safety issues
- Attends all company safety training
- Helps others create maintenance safety work-orders
- Participates in weekly department safety walk and follows up on any open issues
- Participates monthly in company safety walk
- Audits peers for PPE policy compliance weekly
- Is the go-to person for others whenever a safety issue arises
- Has a passion for safety and safety improvement
- Participates in the company safety improvement and safety observation programs

Figure 9.1   Safety contact job description.

next section, can be written and posted when filling these or other safety positions.

Other Team Building Opportunities

- ▪ **Safety contact subteam:** Designate or select a safety contact from each department or work cell. They can function as a subteam of the steering team. Their activities could include departmental safety audits and safety communication. Figure 9.1 gives you a broader overview of a safety contact's possible contributions. This is yet another way to develop a safety culture since you will have involved more people directly in the safety program.
- ▪ **First responder subteam:** When an injury or accident occurs in a facility, there should be a plan on how the subteam is to react rather than hoping someone else reacts appropriately. By forming a team of caring first responders, providing training in first aid and the basic first aid materials and equipment they will respond appropriately when an emergency arises. Through their planned response to stressful,

unplanned injury situations, they bring comfort to both the injured and the management team. The population of every plant, which is a cross section of society in general, has some very caring individuals who would love to fill this type of role. Building this subteam sends a cultural message to all that management cares about people.

■ **PPE (personal protective equipment) subteam:** This team can ensure that PPE requirements are defined for the different tasks within the plant and then conduct audits for compliance. When new equipment or processes are brought into the plant, they can assess the PPE required.

■ **Noise abatement subteam:** Annual noise level audits, annual hearing testing, and defining the need for sound enclosures would fall within the scope of this team.

■ **Lockout/tagout (LOTO) subteam:** Annual training and audits, selecting the required LOTO devices, and the creation and placement of visual LOTO procedures on all equipment would be the responsibility of the team.

■ **Accident/incident investigation subteam:** This team is accountable for the continuous improvement of safety by using every incident and accident as a vehicle to get to root causes and corrective actions. The team leader must have strong lean and interpersonal skills in order to gain and build trust during the investigations.

■ **Ergonomics subteam:** If a business has not addressed the causes of soft tissue injuries, forming and training a small subteam makes business sense. These ergonomics-related injuries are a major contributor to injury totals and workers compensation claims. Once provided with some basic ergonomic risk assessment training, a team can proactively identify a list of work areas that have inherent risks. Then using safety kaizen blitz events, they can engage others in the reduction of the soft tissue injury risks.

■ **Annual safety system audit subteam:** An audit of the entire safety system or program ought to be conducted annually in the same way the company's financial and quality systems are audited. This should be a management review with some senior leadership and the safety team leadership both involved in conducting the audit.

■ **Safety promotion subteam:** This subteam's objective is the raising of safety awareness. They would audit and approve safety improvements and track and support the JSA (job safety analysis) efforts. They could also collect the safety observation forms and hold a drawing to select a few winners at all company meetings. Rewards earned through these three employee engagement programs would be purchased and distributed by this team.

Obviously many people can be engaged in a safety program with the use of subteams. If a company does not want to develop and engage their employees in safety improvement, they can hire a safety director to accomplish all that is noted above. A single lean champion cannot build a lean culture nor can a safety director build world-class safety alone. It will take a team or a committee. The team can be composed of new volunteers and a few staff members who are committed to developing the volunteers. Another option is that the committee can be composed of some current staff members who will tell others they were "voluntold" by their bosses to be part of the safety committee. Who is going to get your plant to world-class safety? I vote for the team.

Recordkeeping

Along with any safety program comes the onerous responsibility of recordkeeping. Many of the records are maintained to prove compliance with OSHA-mandated programs while others can be internal documents that guide and manage your internal safety program: OSHA injury tracking, daily inspection records for hoists/cranes and forklifts, training records for required annual safety training, MSD (material safety data) sheets, HMIS (hazardous material identification system) program, fire (tornado, flood, hurricane) evacuation guidelines, safety program outline documents, safety rules, internal safety metrics, safety incentive program outlines, team meeting minutes, and the team task list, to name a few. Depending on the business, this list can grow substantially. For instance, food and pharmaceutical companies must have full-time staff just to maintain all of the records required to ensure a safe supply of food and drugs to the consuming public.

Lean Approach to Recordkeeping

Lean cannot change the mandated requirements to maintain safety records. They are a required audit trail to prove compliance during audits and legal proceedings resulting from injury claims. Lean's influence on recordkeeping is the added requirement to PDCA (plan-do-check-act cycle covered in Chapter 4), or annually review some program documents. For instance, safety program outline documents, safety rules, and evacuation plans all can benefit from an annual review to guarantee the information is current. The safety program documents describe your safety program and must be continually updated to reflect the continuous improvement of the program. To clarify this important continuous improvement activity, let's review a couple of recordkeeping categories.

Safety Program Outline Documents

These documents can be the heart and soul of your safety program because they describe the components that make up a large part of the total program. These are overview documents that describe the intent and scope of the program's components. Activities and actions will cascade down from each of these outline documents. For instance, all plants should have a PPE component of their safety program. The outline document will, from the skyscraper view, talk about the intent of protecting employees from the regular hazards posed by the materials they handle and the equipment they operate. From the steeple view, it will note that a PPE subteam will ensure all cells or departments are inspected and the PPE requirements will be posted on the cell's bulletin board. From the plant floor view, there may be a list of approved personal protective equipment, such as gloves, safety glasses with side shields, safety shoes, earplugs, or other hearing protection devices, that is attached to the program document. An annual review of the PPE program documents might discover that a type of glove listed on the list of approved personal protective equipment contributed to an injury and is no longer purchased or offered to the workforce. This annual review step is the "check" step in the PDCA cycle. The action step is, of course, to remove the gloves from the list of approved PPE items. Figure 9.2 lists some other possible safety program documents that overview important segments of the safety program. The items listed will vary depending on the type of business, but the "lean learning" is that nothing is or should be stagnant in a safety program. Because of this, a regular review of the safety program overview documents is a necessary continuous improvement activity.

Safety Education

Safety education occurs at many levels and involves different people. Much of it is conducted on the job. New employees are hired, given a company orientation, which may include some safety topics, and then they are turned over to the supervisor from the department where they will be working. The

Safety Program Overview Document List

Program	Owner	Annual Review	Notes
PPE	PPE subteam	9/31/09	
LOTO	LOTO subteam	6/1/09	
HMIS	Manufacturing engineer	7/1/09	
MSDS	Purchasing manager	7/1/09	09 plan to electronically store and retrieve
Blood Borne Pathogens	HR	5/1/09	
Safety Walks	Safety contact subteam	11/1/10	
First Responders	HR	5/1/09	First aid training schedule in Q3
Accident Investigation	Plant manager	5/1/10	
Annual Program Audit	Plant manager	12/1/09	Include safety team leader and HR rep on audit team
Evacuation Plans	HR	1/30/10	
Safety Promotion	Promotion subteam	8/1/10	

Figure 9.2   Safety program overview document list.

supervisor may also cover some safety topics and then turn the employee over to an experienced operator. This is when the real training, and the identification of safety hazards associated with the job, will begin. After the on-the-job safety training, and for the rest of an employee's career, most all of the safety training he/she will receive is mandated. Annual LOTO and HAZCOM (hazard communications) are two that come to mind.

Lean Approach to Safety Education

The opportunities to expand safety training beyond what is mandated are many. Here is a list of ideas.

■ Attending trade shows that focus on workplace safety
■ Visiting other plants to benchmark safety practices

- Participation on a safety subteam or become a member of the safety team
- Safety kaizen blitz team participation
- In house training on lifting, ergonomics, guarding, or other relevant topics
- Attend safety workshops offered by the National Safety Council or other safety organizations
- Participate on JSA teams

   All of these opportunities are outside the scope of one's regular job. This higher level of learning and education will only take place in a culture where employees are seen as an asset. That is the lean safety connection. If a company is not serious about lean, they are not going to be serious about attaining world-class safety. They will want their workers to stay on the job and produce parts rather than grow through involvement in safety education and the application of what they have learned to improve safety.

Safety Program Activity Management

The recordkeeping and task management required to keep a safety program on track and in compliance are often assigned to one individual. Most tasks and records are related to OSHA injury tracking and reporting, internal accident reporting and investigation forms, and the medical and worker's compensation records when injuries occur. The HR department is often the location for this activity. They manage the paper and bird-dog others to submit the documents required to maintain accurate records. The size of the organization will often determine how many people are directly involved. If a company is large enough, they may hire a safety manager or director who assumes these responsibilities.

Lean Approach to Safety Program Activity Management

A company striving to develop a world-class safety program requires a different process to maintain and keep track of all of the safety program documents and activities. Earlier I talked about a safety team that was supported by some key subteams. Both the teams and their broad range of activities can be managed through the use of a well thought out task list. It can be utilized to track current open tasks and to plan the year's significant activities, such as annual training and program annual document reviews. Figure 9.3 is an example of such a team task list.

Company Safety Team Task List

Open Tasks		% Comp	Due Date	Responsible
Open Tasks		**% Comp**	**Due Date**	**Responsible**
1	Register safety team for NSC event		6/22/10	Mel
2	Conduct guarding audit – Cell #5		4/30/10	Gary
Open Accident/Incident Reports				
1	15-09 Laceration – Tool Room	30%	12/15/09	Ray
2	22-09 Forklift damage - Receiving	20%	11/12/09	Sue
3	3-10 Fall in parking lot		2/15/10	Will
Annual Program Document Review				
1	Evacuation Plan	10%	2/1/10	Krista
2	Ergonomics Program		7/1/10	Barb
Equipment Purchases or Moves				
1	New CNC Machining Center	75%	3/1/10	
2	Relocated shipping racking	50%	4/1/10	Jim
Annual Responsibilities				
1	OSHA Forklift training		8/1/11	Greg
2	LOTO Training		7/1/10	Curt
Subteam Open Issues				
1	PPE – define PPE for new work cell	50%	1/30/10	Jeff
2	Promotions team – Audit safety improvements	10%	12/31/09	Xavier
Safety Team Budget				
1	Q4 Management review		12/28/09	Fred
Performance Measurement Review				
1	1 year no LTA	75%	2/11/10	Mark
2	Injury costs (less than 2008)	65%	12/31/09	Cyndi
3	60 JSAs completed in 2009	80%	12/31/09	Elaine
2009 Accomplishments				
1	6 months no LTA	100%	8/12/09	
2	Benchmark visit to ABC Company		7/13/09	
3	Made 2009 safety improvement goal	100%	11/14/09	

Figure 9.3   Example of a team task list.

This is the document that should be used by the safety team leader to conduct the biweekly safety meeting. Updates are given only on the tasks that are due or past due. New tasks are added and any open issues can be discussed. In the past, I sat through safety meetings that lasted from one to two hours and ended with no tangible outcomes, such as a task list. They were filled with complaining, debate, and disagreement. All of that ended when the safety team leader was coached and mentored on how to conduct effective meetings using a task list. After this change, the safety meetings were short, productive, and professionally lead. A good task list is one of the keys to effective safety program management. Of course, all of the OSHA-required recordkeeping responsibilities still exist. HR or the safety manager still owns that responsibility. In theory, they should have more time to ensure compliance if the safety team and subteams have taken ownership of other safety tasks for which they may have been responsible in the past.

Safety Rule Definition

I have visited and toured plants where the employees were not wearing safety glasses when I thought they should have been. Obviously, this is possible because many of the safety rules in plants are internal rules set by management. I most often see this lax approach to safety in smaller firms where they may not want to add additional safety requirements for fear of upsetting the workforce. My guess is they live in their own world, do not benchmark, or visit other plants and probably do not belong to any organizations that might challenge their lackadaisical approach to safety. Many safety rules that are defined by OSHA requirements, for instance, punch press guarding, are very clear and not open to debate. Others, such as wearing safety glasses with or without side shields, the decision to allow or not allow the wearing of jewelry, or safety shoes with or without metatarsal guards are often decided at the plant level.

Lean Approach to Safety Rule Definition

Companies on the journey to world-class safety use the lean approach of benchmarking other's practices to determine the best in class safety rules. Their concern is not how the employees may feel about the added safety rules, but rather a deep concern for the safety of their employees. As a result, they do not wait for mandated changes or changes driven by an accident investigation. Their safety rules changes are guided by a researched understanding of risks and safety. If there is a remote chance that someone could receive an eye injury, everyone will be required to wear safety glasses with side shields. Some will grouse about the change, but what a great time to discuss how leadership really cares about their safety and doesn't want to chance an employee injury. Existing safety rules offer an opportunity to use the PDCA lean tool. A great special meeting topic would be the review (the check step) and validation (act step) that the rules are still appropriate as listed. If not, the meeting attendees (HR, safety team, and management) should come to consensus on removal, modification, or a re-write (plan and do steps). Safety rules definition, like continuous improvement, is never completed. New equipment and new processes, installed in a world-class safety culture facility, trigger a PDCA review of existing rules and the research of any required rules changes or additions to ensure worker safety. If I worked there, that is exactly what I would want to happen. That is a lean safety culture.

Safety Communications

Safety may only be discussed when it has to be, for instance, during mandated safety training or when an accident takes place. If safety communications are seen as unimportant, or not given focus, then the employee base will view safety in the same light. This minimalist approach will lead to little safety improvement and possible large safety costs.

Lean-Influenced Safety Communications

The most important element of any effort to engage and empower the workforce, and drive a company's culture toward world-class lean or safety, is communications. Effective communications are critical because it is through communications that you can gain trust. And only when you gain trust will the culture shift toward the intended direction. The definition of empowerment that I most often use is: "the downward flow of information." Therefore, what information can be provided to guide a so-so safety program, with minimal employee involvement, into one that other plants will want to benchmark? Some of the topics listed below were already mentioned elsewhere in the book, but to fully understand the depth of safety communications available, I will restate your options.

- Biweekly safety meetings with an employee-based safety team
- Safety contacts who lead the discussion on safety topics during their weekly team meetings

- Safety-focused kaizen blitz events with wrap-up meetings to celebrate safety gains
- Teams conducting JSA (job safety analysis) events
- Recognition of all safety target and goal accomplishments at all company meetings
- Proactive safety improvement programs that collect and reward the implemented safety improvements of all employees
- Safety training beyond mandated training for the safety team and others, such as department safety contacts
- Attendance by hourly employees at safety tradeshows, benchmarking visits to other plants, and external safety training events
- Hosting safety events in your facility to allow those involved to showcase their efforts
- Applying for safety awards to benchmark your program against others
- A promotions subteam that provides promotional give-away items emblazoned with the company name and "Safety First"
- Awarding time off (always everyone's favorite reward) for the attainment of a milestone safety goal
- A world-class lean facility, which was attained by focusing on lean safety, being used as a sales and marketing tool
- An annual safety award presented to someone who has made a real difference in your safety program
- A safety newsletter featuring training materials and internal safety programs general interest articles
- A safety team bulletin board located in a prominent location; add metric and other safety team related materials

Change programs intended to impact and redirect the culture of a business can only benefit from increased communications. Call on the creative people in your company to build upon this list of possibilities.

Quick Guide: Roadmap to World-Class Safety

- Your ability to impact people and, therefore, the cultural side of safety will determine your success.
- A safety policy statement should include statements about "management involvement" rather than the typical "management commitment."
- Define management safety-related "standard work" activities that will publically demonstrate their safety commitment.

- Establish a safety team that will work cooperatively toward world-class safety.
- Develop subteams and subteam leaders to manage the parts of the safety program.
- Write safety program outline documents.
- Provide safety education beyond the norm.
- Use a safety program management document to conduct safety team meetings and monitor the required tasks for program success.
- PDCA existing safety rules.
- Provide safety communications in many new and different forms.

Chapter 10

Safety Standard Work

Foundation of Continuous Improvement

Standards are the baseline from which all continuous improvement activities can be measured. An example I often use is that of an order picker in a Toyota distribution center. I am not sure if the facts I use are correct, but that is inconsequential because what is important is the concept. Standard work for order pickers is 12 picks every 15 minutes. If in any 15-minute period they do not make their standard, they have to signal their supervisor. The supervisor immediately responds and approaches the order picker not to ask why he did not work to standard, but instead to ask what process problem, or problems, prevented him from making standard. Was the inventory count incorrect? Was the inventory in the wrong location? These and other questions will be asked until the problem is solved. This short, simple example demonstrates three distinct principles of lean: respect for people, standard work, and a continuous improvement culture. Having defined standards that everyone understands enables the other two principles. Unlike this example, the actual workday of many supervisors is one spent firefighting rather than focusing on continual improvement. Moving from one problem to the next fills their days because standards are unknown or not communicated.

Recently in the lean community, two different methodologies related to standard work have been receiving focus. One is the Training Within Industry (TWI) and the other is manager standard work. TWI's three programs, job relations, job instruction, and job methods, work together to help supervisors

- Build and maintain positive employee relations.
- Train workers to quickly remember how to perform jobs correctly, safely, and conscientiously.
- Improve the ways in which jobs are done.[1]

The second approach, standard work for managers, is a powerful method that can help drive the actions required to build a lean culture. For example, you can spend months training everyone in the principles of 5S (sort, set in order, shine, standardize, sustain) and then a few more guiding them as they implement the program throughout a facility. 5S will only become cultural if an audit method, the last S, which is sustain, is clearly defined, and adhered to. This is where standard work for managers can help. If on every supervisor's Outlook calendar they were scheduled to conduct a one-hour 5S audit of their department each and every month, you would have a standard work process in place. The audit step for this standard process would be the submission of the audit form to their manager. Failure to submit the audit forms on time would be the equivalent of the order picker signaling his supervisor when he could not work to standard. Can this lean method used to drive cultural change be used to drive safety culture change? Yes. Imagine a safety walk appointment on a supervisor's Outlook calendar the first Monday of every month from 1:00 until 2:00 p.m. That would be a safety standard work schedule. To extend this concept further, the following are some common position titles along with some possible safety standard work activities they could perform to help build world-class safety.

- Plant manager or general manager:
  - Once a week take a safety metric walk to review the team's safety metrics posted on their team boards.
  - Attend the biweekly employee safety team meeting.
  - Once a quarter participate in a different work cell safety walk.
  - Once a year take the safety team out to lunch.
  - Attend every safety kaizen blitz wrap-up presentation meeting.
  - Make safety the first topic of every meeting you lead.
- Frontline supervisors:
  - Complete two safety observations every week. Catch someone working safety and recognize him for his safe behavior.
  - Every morning at the cell flash meeting update the safety calendar with a green safety cross if the day before was an injury-free day.

- Conduct a monthly 5S audit of the cell or department.
- At weekly team meetings, review team safety metric performance against target. Metrics could include number of safety observations, safety improvements, or safety JSAs completed.

■ HR safety representative:
- Maintain and update company safety metric board.
- Attend biweekly employee safety meeting.

■ Department or cell safety contact:
- Review weekly safety focus every morning at cell/department flash meetings.
- Conduct a weekly safety walk within cell or department.
- Conduct a quick, daily PPE audit in your work area.

■ Lean champion:
- Facilitate a defined number of safety kaizen blitz events annually.
- Lead all accident investigations using lean tools to get to root cause and corrective actions.

■ Employee safety team leader:
- Walk the plant for a half hour every day to provide the opportunity for all to raise safety issues.
- Lead the biweekly employee safety team meeting.

■ Work cell or departmental teams:
- Target a number of safety improvements, safety observations, and JSA that will be completed monthly. Review metric results every week at team meeting,

These are but a few examples of safety tasks that can, if scheduled, acted upon, and then reported on, develop a safety standard work culture within a facility. This safety standard work culture will then drive the continual improvement of safety.

TWI has its roots in the hands-on, massive work training effort required during World War II. It has taken on new life with the lean community's understanding of the requirement for baseline standards against which to measure your lean gains. This old concept of standards for the methods used for hands-on work and the training to perform those tasks has been extrapolated into standard work for managers—a relatively new concept. The value of standard work for managers is the early-on requirement for managers to sit down with their next level manager and decide what is important. This discussion will force the weeding out of some tasks and a consensus on the critical tasks that should have standard work methods put in place to ensure they are addressed. Standard work methods are applicable to the monitoring of safety, quality, or production processes. Standard work can be the audit step that ensures success of any large-scale change effort. Many companies who are seriously pursuing lean are adopting this new thinking. So rather than continuously coping with problems, managers at all levels, are devising and then following plans that allow them to focus on continuously improving their operation.

Quick Guide: Standard Work for Managers

- Standards are the baseline from which all continuous improvement activities can be measured.
- Standards are often unknown or not communicated.
- TWI and standard work for managers both support the cultural change required for lean or safety implementation.
- When implementing standard work for managers, an early-on requirement for managers is to sit down with their next level manager and decide what is important.
- Standard work can be the audit step that ensures success of any large-scale change effort.

Endnote

1. The TWI Institute overview; http://twi-institute.com/

Chapter 11

Safety Metrics

What You Measure Makes a Difference

Measurements or metrics are critical to a lean implementation. Without a starting point, it is impossible to know your progress. Lean is all about change and change can be …no, it is … difficult for everyone. Differing views will often be presented during an attempted lean implementation by two factions: (1) those trying to hang onto the old way of doing things and (2) the change agents who are trying to turn their world upside down. If someone has spent a career relying on inventory to protect him/her against all the uncertainties of manufacturing, he/she is not going to be very open to the suggestion to do without safety stock. These factional debates can lead to friction, confrontation, subversion, and a good bit of stubbornness. Yet, protest for protest's sake often gets one nowhere. Think of the environmental movement from decades past. Protesters would chain themselves to trees or equipment to prevent the cutting of timber or engage in some other acts to make their point. On a grand scale, I do not think they accomplished what they hoped via their emotional and confrontational style of protesting. Yet, some of these protesters, or at least environmentalists, finally figured out that facts, and some high profile marketing, are the key to driving environmental change. Today new terms, such as sustainability, carbon footprint, and global warming have almost everyone's attention. The "green" protest movement is becoming mainstream thanks to former vice president Al Gore and others who have taken it to center stage. My point is that they used facts, not protest, to get the public's attention. Lean leaders, change agents all, must also use facts if they are to convince others to join in the change. Not just facts on paper, but factual evidence observed on the shop floor that can convince others a one-piece pull cell will efficiently serve the customer base without excess inventory. Facts are the heart of continuous improvement or lean. Using facts, rather than emotion, will always help to bridge a gap between people. How can facts, in the form of metrics, help you attain world-class safety?

Walk into any plant and you should be able to find a company metric board that contains a safety-related measurement or two. Most frequent might be some document that tracks the number of days without a "lost time accident" (LTA). LTAs, noted by the terminology alone, are more serious than others since someone was injured and, as a result, unable to report to work. These are the injuries that result in worker's compensation claims and payments. This category of injury is almost universally measured because it is required and because they are costly—both to the injured and to the business. Many companies set a very visible goal to get to one year with no LTAs. A downside of attaining that goal is the possible complacency that may follow that accomplishment. Often a flood of injures are reported soon after goal attainment because individuals with bad backs or other soft tissue problems suffered rather than report the condition and put a stop to the record. Companies are justifiably proud when they attain a year or sometimes multiple years without a LTA. No LTAs in a period is a reason to have another safety banner printed and hung. It is a skyscraper metric that rolls up and displays the results of the entire safety program. A no LTA metric is suitable for front hall display and board meeting reviews, but it is not time-relevant to the hourly worker. A lean thinking approach is not about changing this metric, but rather to define proactive programs with metrics, at a shop floor level, in order to arrive at multiple years without a LTA. Continual improvement, an integral part of a world-class safety program, minimizes the chance of complacency after the LTA goal is arrived at. In a lean culture, the no LTA accomplishment is a result of continuous improvement activities rather than just everyone wishing and hoping and trying to be safe for a year. It is the result of proactive efforts. There is no let down, only more opportunities to take safety to new levels.

Lean Approach to Safety Metrics

■ **Audit scores:** Many parts of a safety program have an audit component. After the first audit of a work cell or department, a benchmark number has been set that can be used to establish a goal for continuous improvement. Audit scores issued for 5S (sort, set in order, shine, standardize, sustain), department safety walks, and PPE (personal protective equipment) compliance can all be used to drive ongoing safety improvement. Those same cellular or department audit scores can be rolled up into a plant score. Targets can be set and celebrations held when they are met.

▪ **Number of incidents reported:** Conventional safety program thinking views an increase in incidents (near misses) as a precursor to increased accidents. There is also a general concern that setting targets for the number of incidents or even accidents to drive down the number may cause individuals to hide injuries and incidents. If that happens, there is "fear in the workplace" or, to say it another way, there is a lack of trust in those investigating the accidents. Businesses with a past practice of relying on discipline when violations occurred will drive reporting underground. As stated in Chapter 7, under the section on incident investigations, lean thinkers see incidents as opportunities to improve safety. Incidents tracking could be seen as a positive metric for it demonstrates the level of trust in the investigation process. Trust that root cause and corrective actions are the goal, not finding the individual responsible so that discipline can be issued.

▪ **Number of safety improvements:** Proactive safety improvements that help to reduce the risk of injury are the objective of a formal safety improvement program. As a metric, the number of safety improvements can be tracked at the work cell, department, business unit, or company level. It is a time-relevant metric because everyone can impact it every day through their identification of safety improvements. Individuals implement and then submit their improvements on a form to the safety promotion subteam for approval. If approved, the originator receives some level of reward. This program engages everyone as individuals or as a member of a work group in the continual improvement of safety. If safety is a component of a company's annual employee review document, a targeted number of safety improvements could be set as a target for each and every individual. If pay rates are linked to those annual review scores, there is now a strong connection between a reward, hourly pay rate, and an expected behavior—focusing on safety improvement.

▪ **Number of safety observations:** I mentioned this program earlier. It is yet another proactive way to engage all employees by asking them to observe and recognize their peers for safe behaviors. A target can be set at the individual, department, or company level. Someone in our plant created a "safety wheel of fortune" that was used to award prizes during quarterly town hall meetings for the safety observation program. Anyone who had submitted a safety observation form was entered into a drawing. If your name was drawn, you spun the wheel and collected cash or company logo safety-related gifts. It was a fun way to celebrate safety. Since this program focuses on employees' safety behaviors, it supports a move toward behavioral-based safety programs.

■ **Number of job safety analyses (JSAs) completed:** Job safety analysis exercises are minikaizen events. By targeting a number of them in a period, e.g., 20 JSA events in 2010, a company will guarantee employee engagement and improvement of the safety program. A side benefit of JSAs and safety kaizen blitz events is that the team members identify numerous safety improvements through their efforts. In our facility, those same people would submit the safety improvements for approval, based on the program described in the last section, and receive their rewards.

Some may have already recognized that all of the above possible metrics are proactive ways to positively impact safety in a plant. None of them are required or mandated. All of them help the employee base become more safety aware and better able to identify safety hazards. Any one of these will shift a safety culture toward world class.

Compliance with OSHA regulations will not prevent all potential hazards in a plant. Tracking the OSHA-mandated metrics will not drive safety improvement. Both of these efforts are what companies have to do. This book was written to encourage individuals and the companies where they work to go beyond compliance. Most businesses today are somewhat lean savvy and have someone on board championing these lean efforts. This individual, armed with lean tools, can engage others and then together make a real safety difference. And safety is the no. 1 priority, right?

Quick Guide: Metrics

■ Without a starting point, it is impossible to know your progress.
■ Define proactive safety improvement programs with metrics.
■ Use safety audit scores to set improvement targets.
■ Use incident investigations to drive safety improvements and track them.

- Build a safety improvement program that all can participate in and set a targeted annual number of improvements for the company.
- Start a safety observation program that allows everyone to recognize safe behaviors. Set an annual target.
- Target a number of JSAs to be completed in a period.

Chapter 12

Conclusion

Much like lean practitioners who fuel their passion by challenging processes and growing people, individuals who are passionate about safety recognize that their passion is derived from the opportunity to interact with, develop, and protect people. People engagement is the common denominator in the necessarily painstakingly slow journey toward world-class lean or world-class safety. It is difficult to change, grow, or develop a new business culture quickly. It is a multiyear journey. Companies that elect to follow the lean safety path to engage their staff can accomplish both safety and lean growth concurrently and speed their journey toward world class.

"Safety first" is really just a safety slogan printed on banners and posters. It can accomplish nothing more than remind those who read it about their safety responsibility. In reality, those charged with running a business are not thinking "safety first" throughout their workday. Safety is something they and we must each internalize and only then can we define our own safety banner. Our imaginary safety banner is a compilation of the safety decisions we make each and every day: at work, when driving a car, or when we are involved in anything that has inherent danger. I hope this book helps you define yours and, when completed, it surrounds and protects you, your co-workers, family, and friends because you have chosen to include the words "safety first" on it. Stay safe.

Glossary

There are many great reference books that provide a broader, more in-depth definition for the terms I have included. My intent was to help those in the safety community, who may not be familiar with some common lean terminology, gain a basic understanding while reading this book.

A3: This is another TPS lean tool. This problem-solving, project management methodology merges activities, such as process mapping, problem identification and solving, goal setting, and auditing, and does it all on one piece of paper, front and back.

AME: Association for Manufacturing Excellence. This not-for-profit organization was founded in 1985 and is dedicated to helping companies with continuous improvement and their pursuit of excellence.

Andon: A system used to notify others of equipment or process problems. Often a light pole mounted on operational equipment that has three different colored lights. The lights, which signal machine condition, can be activated manually or via the machine controls.

Benchmarking: A formal or informal method of comparing and sharing business processes and performance metric results in order to drive business improvement.

Continuous flow: A lean term used to describe the goal of eliminating batch production. By moving toward a batch size of "one," many of the wastes associated with a batch production process (storage, movement, inventory) are reduced or eliminated. The safety benefits of continuous flow derived from reduced material handling can also be substantial.

Current state: The current state is "how it is today." When using lean tools like the A3, process maps, value stream mapping, or a kaizen blitz, the facilitator may guide the team to a full understanding of the current process by having them complete a current state map.

Cycle time: The time it takes to complete a discrete step of an operation. All of the steps that comprise a business process can be timed to understand the current total cycle time.

Empowerment: A premeditated effort to drive decision making down the organizational chart. True empowerment requires a downward flow of information with which decisions can be made and a leadership team willing to let go of some responsibility.

Facilitation: A leadership role in which one guides a team of individuals toward an outcome. The facilitator should remain content neutral and accept responsibility for only the process followed to reach the outcome.

5S: A foundation lean tool used by organizations when beginning a lean journey to develop a baseline discipline in the workplace that can then be built upon using other lean tools. It is a workplace organization methodology that engages everyone in the activities required to embrace the 5Ss: sort, set in order, shine, standardize, and sustain.

Future state: The future state can be described as "how it could be." Once a thorough understanding of the current state process has been gained, a facilitator will engage the team in problem solving and improvement idea generation. Based on these ideas for changing the current state, a future state process map will be drawn.

Gemba: A lean term that means the "shop floor." You will often hear lean champions make a statement like, "Let's go to the gemba and get the facts."

Hazmat: Acronym that refers to hazardous material that may be used in the workplace.

HMIS: Hazardous material identification system. Color-coded labels combined with a numerical designation used to identify potentially hazardous materials.

JSA: Job safety analysis is a detailed review of the safety steps related to a particular work task. The steps are recorded on a standard form that can then be used to train individuals in the best practical safe way to perform their work tasks. This is a proactive method that can help to develop a world-class safety culture.

Kaizen blitz: A team-based rapid continuous improvement activity. These hands-on events can last from three to five days. The team is challenged to reach a stretch improvement objective. A facilitator then guides them using a standard process to goal attainment. Because they are learning by doing activities, a kaizen blitz can quickly impact an individual's understanding and acceptance of lean principles.

Kanban: Often described as a signaling system. This lean tool is most often used as replenishment method for raw materials, work in process of finished goods. Its use enables pull production processes that replace the push-based, work order replenishment processes.

Lean: A manufacturing philosophy that reduces the total cycle time between taking a customer order and the shipment by eliminating waste. Lean is based on the Toyota Production System (TPS). Since lean is a process review and improvement activity, it can be applied to all business processes, including safety, and all business types. Application of lean is taking place in medical, military, government, service, process, and manufacturing operations.

Lean safety: The application of both the lean tools and management philosophy to create a continuous improvement safety culture.

Lost time accident: An accident that results in a nonfatal injury that causes the injured to lose time away from work.

LOTO: Lockout/tagout. A safety procedure that helps to ensure that all power sources are turned off and isolated to prevent injury while performing maintenance or repairs on equipment. Standard keyed locks and other energy isolation devices are used to perform the lockout/tagout step.

MSDS: Material safety data sheets. A form supplied by the producer of a product that contains safety information regarding the product and its base ingredients. Information can include health hazards, emergency treatment methods, toxicity, reactivity, storage, disposal methods, and protective equipment required for safe handling. An MSDS should be supplied at the time of purchase of a product and kept on file at the place of use.

National Safety Council: A national organization devoted to: "Educating and influencing people to prevent accidental injury and death." Most states have a NSC chapter.

PDCA: An abbreviation that stands for plan, do, check, and act. It is a four-step problem-solving process typically used in business process improvement.

PPE: The abbreviation for a class of personal protective equipment used to protect an individual from injury or illness while on the job. It can include gloves, earplugs, safety glasses, helmets, goggles, face shields, aprons, safety shoes, and many other devices. Defining and providing the correct PPE for each particular job is an important step in developing an effective safety program.

Process map: Process mapping, or the expanded version, value stream mapping, is used to graphically display the steps in a process. This visual method of process evaluation and improvement engages everyone because it is completed on white boards, flip charts, or by using Post-it® notes arranged on a wall. Both a current state and a future state map are usually completed during lean events.

Process owner: Each defined business process should have an individual assigned to monitor and direct that process. Often when business processes are ineffective or produce poor results, it is due to the fact that process ownership is unclear.

Root cause: The one cause that leads all of the others when conducting an investigation of an outcome. In accident or safety incident investigations, the pursuit of effective corrective actions is dependent on defining the root cause of the accident or incident.

Safety contacts: A volunteer safety representative from each department or work cell in a facility. They are the communication link to the company safety team and the first responder to any safety concerns that come to light in their work area.

Safety improvement: A proactive idea implemented to help prevent injuries from occurring in the future. A safety improvement program is a method that can be used to engage the entire workforce in proactive safety improvement activity.

Safety observation: Someone directly observing someone else while he/she performs a work task safely. A safety observation form can be created and then be used by everyone in an organization to recognize safe behaviors. This formal program can engage everyone in safety and have a cultural impact.

Safety team: A formal team-based structure that distributes responsibility for safety process ownership. Safety team membership requirements can be based on desire or interest rather than title. It can be cross functional because the team can and should be staffed from all areas and all levels of the organization.

Safety walk: A formal safety inspection process. An individual or group of individuals, using a safety check sheet, walks through a department or facility with an eye for safety. Any noncompliance issues or situations are noted on the inspection form.

Standard work: Having work standards will provide a clear and agreed upon understanding of how a job task will be performed and what the expected results should be. This understanding is a building block for a lean culture.

TPS: The Toyota Production System. The management philosophy and system developed at Toyota, which is the foundation of what today is called "lean manufacturing."

Visual factory: A lean methodology used to disseminate operational information to everyone using visual means or methods. If implemented effectively, it can eliminate the need for conversations and computer-generated reports and make needed information "public" or accessible to everyone.

Waste: Anything the customer is unwilling to pay for. All processes can be broken down, or mapped, to yield a series of steps. These steps can then be categorized as value-added or nonvalue-added. The nonvalue-added steps, the ones the customer would not want to pay for, are considered "waste" to a lean thinker.

Index

About the Author

Robert B. Hafey has worked in manufacturing operations and maintenance for 40 years. He began his career with U.S. Steel Corporation, followed by over 20 years at Flexco. For the past 18 years, he has been directly involved in the definition and implementation of continual improvement at Flexco. He has been an AME (Association for Manufacturing Excellence) volunteer for the past 14 years and acquired much of his lean knowledge through this involvement. Hafey has served on the national board of the AME, as well as a regional board president, regional board member, and has been a volunteer member of three AME annual conference teams. This affiliation and others have provided him the opportunity to enhance his lean leadership and facilitation skill set beyond his place of employment. He has presented papers on the topic of continual improvement at conferences sponsored by AME, the Society of Manufacturing Engineers (SME), the National Industrial Belting Association (NIBA), and the Manufacturing Institute, which is located in Manchester, England.

Hafey firmly believes in the e-mail signature tagline he created: "You can continuously cope or you can continuously improve—the choice is yours!" He considers continuous improvement a creative endeavor and he enjoys and finds time for many other varied activities that center on creativity. His creative instincts and an undergraduate degree in professional arts are the foundation that enabled him to master detailed woodcarving, gourmet cooking, and guiding others on their continuous improvement journeys.

Hafey resides in Homer Glen, Illinois, with his wife Sandy. They have three daughters, Liz, Kate, and Colleen.